Texts in Algorithmics
Volume 11

London Algorithmics 2008
Theory and Practice
Dedicated to Maxime Crochemore
on his 60th Birthday

Texts in Algorithmics Series Editor
Costas Iliopoulos   csi@dcs.kcl.ac.uk

London Algorithmics 2008
Theory and Practice
Dedicated to Maxime Crochemore
on his 60th Birthday

Joseph Chan
Jacqueline W. Daykin
and
M. Sohel Rahman
editors

© Individual authors and College Publications 2009. All rights reserved.

ISBN 978-1-904987-97-0

College Publications
Scientific Director: Dov Gabbay
Managing Director: Jane Spurr
Department of Computer Science
King's College London, Strand, London WC2R 2LS, UK

http://www.collegepublications.co.uk

Original cover design by Nafees Uddin Ahmed
Cover produced by Laraine Welch
Printed by Lightning Source, Milton Keynes, UK

Preface

This volume of the Texts in Algorithmics series is a collection of work by the participants and friends of London Stringology Days (LSD) and London Algorithmic Workshop (LAW) 2008, sponsored by the Algorithm Design Group and the Department of Computer Science of King's College London. The momentum and prestige of these annual meetings continue to grow. Indeed, 2008 saw a record number of attendees from both national and international affiliations, ranging from research students through to experienced practitioners. The form of this volume is that of a special issue, focussing on core computer science theory along with bringing that theory into the real world of computing via practical implementation. Further, the diverse applications explored underpin the importance of automata, combinatorial algorithmics and stringology.

In this special issue we present an exciting range of current research. The ubiquitous finite automaton is investigated for finding repetitions in text, simulating nondeterminism via a deterministic state cache, and efficiently computing the edit-distance between a string and a weighted automaton. Variants of sequence alignment tasks are demonstrated by transposition comparison networks. Musical studies concern automatically identifying salient polyphonic patterns, and song classification according to rhythm. Experimentation of heuristics for the NP-hard multidimensional assignment problem is assessed, while lossless image compression is evaluated on parallel architectures. Overlap graphs are constructed and simulated for computing the probability of occurrences of words in text, and an independent proof is given for estimating microruns in a string. Moreover, efficient algorithms are designed for determining noise induced errors in magnetic resonance diffusion tensor imaging.

We would like to thank the authors for these valued contributions along with the referees for generously providing such a rigorous reviewing process, who together with the organizers and participants of LSD and LAW mark the success of these events.

It is a great pleasure to dedicate this special issue to our esteemed colleague, Professor Maxime Crochemore, on the occasion of his 60th birthday. Maxime has been instrumental in establishing algorithmic stringology as a vibrant and fundamental branch of computer science. Indeed, it is testimony to Maxime's popularity as a colleague, and stature as a scientist, that such high calibre papers have contributed to the success of this special issue. We are further delighted in honouring Maxime here with the introduction of the notion of a *Crochemore Set*.

Joseph Chan
Jacqueline W. Daykin
M. Sohel Rahman

Table of Contents

A Different Proof of the Crochemore-Ilie Lemma Concerning Microruns[1]

FRANTISEK FRANEK AND JAN HOLUB

ABSTRACT. We present a different computational proof of the estimate of the number of microruns established in a recent Crochemore-Ilie paper. The original proof in the Crochemore-Ilie paper relies on computational means, and thus our proof provides an independent verification of the fact. We also introduce the notion of an R-cover that is essential to our approach. The hope is that a further analysis of R-covers will lead to a non-computational proof of the upper bound of the number of microruns.

1 Introduction

An important structural characteristic of a string over an alphabet is its periodicity. Repetitions (tandem repeats) have always been in the focus of the research into periodicities. The notion of runs captures maximal repetitions which themselves are not repetitions and allows for a succinct notation ([10]). Even though it had been known that there could be $O(n \log n)$ repetitions in a string of length n ([1]), it was shown in 1997 by Ilioupoulos, Moore, and Smyth that number of runs in Fibonnaci strings is linear ([7]). In 2000, Kolpakov and Kucherov proved that number of runs was linear in the length of the input string ([8]). Their proof was existential and thus did not specify the constants of linearity. The behaviour of the **maxrun function** $\rho(n) = \max\{r(x) \mid all\ strings\ x\ of\ length\ n\}$, where $r(x)$ denotes the number of runs in a string x, became an interest of study to many. In several papers (e.g. [4], [11], [3]) several conjectures about $\rho(n)$ were put forth:

(1) $\rho(n) < n$,

(2) $\lim_{|x| \to \infty} \frac{\rho(x)}{|x|} = \frac{3}{1+\sqrt{5}}$,

[1]Supported in part by a grant from the Natural Sciences & Engineering Research Council of Canada, a grant from the Ministry of Education, Youth and Sports of Czech Republic, and a grant from the Czech Science Foundation.

(3) $\rho(n+1) \le \rho(n)+2$,

(4) for any n, there is a cube-free binary string $\boldsymbol{x}$ so that $\boldsymbol{r}(\boldsymbol{x}) = \boldsymbol{\rho}(\boldsymbol{x})$.

[4] introduced a construction of an increasing sequence $\{\boldsymbol{x}_n : n < \infty\}$ of binary strings "rich in runs" so that $\lim_{n\to\infty} \frac{\boldsymbol{r}(\boldsymbol{x_n})}{|\boldsymbol{x_n}|} = \alpha$, where $\alpha = \frac{3}{1+\sqrt{5}} \approx 0.927$. The technique was used by Franek and Yang to provide an asymptotic lower bound for $\boldsymbol{\rho}(n)$ ([5]). This proof was significantly simplified by Giraud ([6]). Just recently, [9] improved the lower bound, falsifying the conjecture (2). The current value of the lower bound 0.944565 (not published yet) can be found at the web site of one of the authors at

http://www.shino.ecei.tohoku.ac.jp/runs/

An explicit upper bound $6.3n$ was first given by Rytter in 2006 and immediately improved by him to $5n$ (see [12]), later improved more to $3.44n$. Crochemore and Ilie ([2]) lowered the upper bound to $1.6n$ using a different method. The current value of the upper bound standing at $1.048n$ (not published yet) can be found at the web site of Ilie at

http://www.csd.uwo.ca/~ilie/runs.html

The Crochemore-Ilie approach relies in two estimates: the first is an estimate of the number of so-called δ-runs, and the other is an estimate of the number of microruns, i.e. runs with period ≤ 9. The first estimate is proven in the paper and states that in average, each interval of length δ contains at most one center of δ-run. The estimate of the number of microruns (Lemma 2 in the paper) states that *the number of microruns is bounded by the length of the string*. As a sketch of the proof, one of 512 different cases is analyzed. The supposedly complete and exhaustive list of all cases was generated using computer. So, the estimate of the number of microruns is established using computational means.

Since it is important for computational results to have independent verification, we present a totally different approach that establishes by computational means the estimate of the number of microruns. We also introduce a notion of R-cover that is essential to our approach. The hope is that a further analysis of R-covers will lead to a non-computational proof of the number of microruns.

2   Preliminaries and definitions

Definition 1. $\boldsymbol{x}[s..(s+ep+t)]$ is a **run** in a string $\boldsymbol{x}[1..n]$
if $\boldsymbol{x}[s..(s+p-1)] = \boldsymbol{x}[(s+p)..(s+2p-1)] = \cdots$
$= \boldsymbol{x}[(s+(e-1)p)..(s+ep-1)]$ and
$\boldsymbol{x}[(s+(e-1)p)..(s+(e-1)p+t)] = \boldsymbol{x}[(s+ep)..(s+ep+t)]$,

where $0 \le s < n$ is the **starting position** of the run, $1 \le p < n$ is the **period** of the run, $e \ge 2$ is the **exponent** (or **power**) of the run, and $0 \le t < p$ is the **tail** of the run. Moreover, it is required that either $s = 0$ or that $x[s-1] \ne x[s+2p-1]$ (in simple terms it means that it is a leftmost repetition) and that $x[s+(ep)+t+1] \ne x[s+(e+1)p+t+1]$ (in simple terms it means that the tail cannot be extended to the right). It is also required, that the **generator** of the run, $x[s..(s+p-1)]$ is **primitive**, i.e. not a repetition itself.

$x[s..(s+2p-1)]$ is referred to as the **leftmost square** of the run, and $x[(s+(e-2)p+t+1)..(s+ep+t)]$ as the **rightmost square** of the run (*for illustration see Figure 1*).

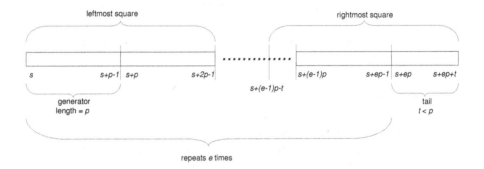

Figure 1. Illustration of a run.

Note that each run can be uniquely encoded by the four-tuple (s, p, e, t).

The core of a run is an auxiliary notion used to construct R-covers (see Lemma 1). Intuitively, it is a set of positions which "destroys" the run if we split the run there into two parts.

We employ the convention that splitting a string $x[1..n]$ in position k means breaking it into $x[1..k]$ and $x[k+1..n]$.

For instance, a run **aaa** cannot be destroyed by splitting: a|aa preserves a run **aa** from the original run **aaa**, aa|a does likewise, and thus it has an empty core. On the other hand, **ababab** can be split into aba|bab and the run is destroyed, so position 3 is in the core, while 2 is not since ab|abab preserves a run **abab** from the original run **ababab** (*for illustration using a more complex run* **abaabaa** *see Figure 2*).

Definition 2. The **core** of a run $r = (s, p, e, t)$ is the set of positions where the leftmost and the rightmost squares of the run overlap less the last index,

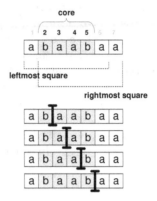

Figure 2. Core of a run and the "destruction" of the run by splitting it in a core position.

or more precisely

$$\{i : s \leq i < s+2p-1 \ \& \ s+(e-1)p-t \leq i < s+ep+t\}.$$

Note: Any run with power ≥ 4 has an empty core regardless its period, so does the cube with period 1 (**aaa**), in a sense these runs are indestructible by splitting. Cubes with higher periods have non-empty cores, and so do squares. If a square has no tail, the core actually contains all positions (except the last one), that is a maximal core – such run can be destroyed by splitting it anywhere, for instance **abab**: a|bab or ab|ab or aba|b, none of the splitting preserves anything of the run.

It seems intuitively clear that in a string with a maximum number of runs, the runs must be distributed "uniformly" and "densely". The following notion of R-cover is an attempt to describe a dense distribution of runs in a string (*for illustration, see Figure 3*).

Definition 3. A set $\{r_i : 1 \leq i < m\}$ of squares in a string x is an **R-cover** of x if

1. the union of all squares r_i is the whole of x;

2. each square r_i has a primitive generator;

3. each square r_i is leftmost (i.e. cannot be shifted left);

4. the starting position of r_i < the starting position of r_{i+1}, and the end position of r_i < the end position of r_{i+1};

5. for any run r in x, the leftmost square of r is a substring of some r_i.

In the following, for the sake of simplicity, a **microrun** (**microsquare**) indicates a run (square) with period ≤ 9. A **micro-R-cover** is an R-cover consisting of microsquares. $\mu(x)$ denotes the number of microruns in x.

Lemma 1. Let $x = x[1..n]$ be a string. If every $1 \leq i < n$ is in the core of some microrun in x, then there exists a micro-R-cover of x.

Proof. Among all microruns that have 1 in its core, choose the one with the largest period, call it R_1. Set r_1 to the leftmost square of R_1. We proceed by induction.

Assume to have constructed $\{R_1 : 1 \leq m\}$ and $\{r_i : 1 \leq m\}$ such that $\{r_i : 1 \leq m\}$ satisfies 2-5 from Definition 3 and each r_i is the leftmost square of R_i. If $\bigcup_{1 \leq i \leq m} r_i = x$, then condition 1 from Definition 3 is satisfied and $\{r_i : 1 \leq m\}$ is an R-cover and the proof is complete.
Otherwise pick the leftmost position $k \in \{1, \cdots, n-1\}$ that is not covered by $\bigcup_{1 \leq i \leq m} r_i$. Among microruns that have k covered by its leftmost square (at least one such must exists, since there is at least one that has k in its core), choose the leftmost ones, and among those, choose the run with the largest period, it is R_{m+1}. Set r_{m+1} to the leftmost square of R_{m+1}.
Since k is not covered by any r_i, $1 \leq i \leq m$, it is not in the core of any of the microruns R_i, $1 \leq i \leq m$, in fact k is to the right of the core of any R_i, $1 \leq i \leq m$. Since k is either in the core of R_{m+1} or to the left of the core of R_{m+1}, R_{m+1} is distinct from all R_i, $1 \leq i \leq m$.   ∎

The following definition of cut is another auxiliary notion. It allows to carry induction over number of microruns: if a string $x[1..n]$ has a cut k, then $\mu(x[1..n]) \leq k + \mu(x[k+1..n])$.

Definition 4. A position $k < n$ in a string $x[1..n]$ is a cut, if the number of all microruns with starting position $\leq k$ is $\leq k$, and it is a smallest such k.

The following lemma is crucial for our proof, it guarantees that under some conditions, a cut exists.

Lemma 2. Let x be an arbitrary string with a micro-R-cover $\{r_i : 1 \leq i \leq m\}$. Let r_1 have not tail. Let there be another microsquare s with a irreducible (not a repetition) generator of size $<$ the period of r_1, with no tail, and starting at position 1. Further assume that $|x| > 35$. Then x has a cut.

Proof. Note that due to the size of x, $m \geq 2$.
The proof is computational and was carried out by the following steps.

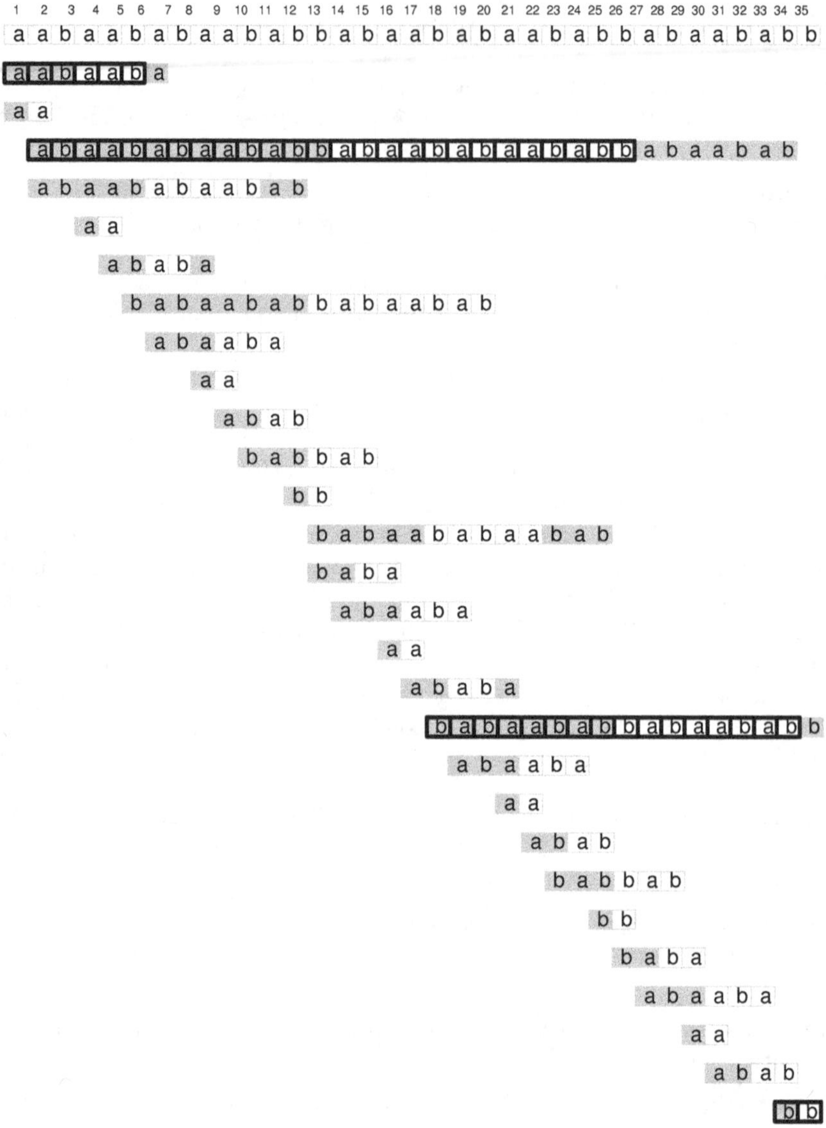

```
 1  2  3  4  5  6  7  8  9 10 11 12 13 14 15 16 17 18 19 20 21 22 23 24 25 26 27 28 29 30 31 32 33 34 35
 a  a  b  a  a  b  a  b  a  a  b  a  b  b  a  b  a  a  b  a  b  a  a  b  a  b  b  a  b  a  a  b  a  b  b
```

a a b a a b a

a a

a b a a b a a b a a b b a b a a b a b a a b a b b a b a a b a b

a b a a b a b a a b a b

a a

a b a b a

b a b a a b a b b a b a a b a b

a b a a b a

a a

a b a b

b a b b a b

b b

b a b a a b a b a a b a b

b a b a

a b a a b a

a a

a b a b a

b a b a a b a b b a b a a b a b b

a b a a b a

a a

a b a b

b a b b a b

b b

b a b a

a b a a b a

a a

a b a b

b b

Figure 3. A string with all its runs, its R-cover is highlighted.

1) s was generated, then it was extended to r_1.

2) The cut k_1 for r_1 was computed (that the cut must exist follows from the fact that $\rho(n) < n$ for all $n \leq 35$).
If r_2 starting position $> k_1$, then k_1 is a cut for $r_1 \cup r_2$ as well. Thus we tried to generate a "bad" r_2 with a starting position $\leq k_1$.
For most r_1 generated, only "good" r_2 could be generated, and so k_1 was the cut for $r_1 \cup r_2$ and thus the cut for $\bigcup_{1 \leq i \leq m} r_i = x$ as well.
In a few cases a "bad" r_2 was generated. The configuration r_1, r_2 was then processed further.

3) The cut k_2 for $r_1 \cup r_2$ was computed (it was always successful).
If r_3 starting position $> k_2$, then k_2 is a cut for $r_1 \cup r_2 \cup r_3$ as well. Thus we tried to generate a "bad" r_3 with a starting position $\leq k_2$. Only "good" r_3 could be generated, and so k_2 was the cut for $r1 \cup r_2 \cup r_3$ and thus the cut for $\bigcup_{1 \leq i \leq m} r_i = x$ as well.

■

3 The main theorem and its proof

Theorem 1. For any string x, $\mu(x) \leq |x|$.

Proof. It is known from various computational results, including the computations carried by the authors of this paper, that $\rho(n) < n$ for all $n \leq 35$, and so $\mu(x) < |x|$ for all strings x of size ≤ 35.
So we can assume that the size of $x[1..n]$ is bigger than 35. We proceed by induction. At each stage, we discuss two cases.

Case 1: *there exists k, $1 \leq k < n$, that is not in the core of any microrun.*
Then $\mu(x[1..n]) \leq \mu(x[1..k]) + \mu(x[k+1..n])$. By the induction hypothesis, $\mu(x[1..n]) \leq \mu(x[1..k]) + \mu(x[k+1..n]) \leq k + (n-k) = n$.

Case 2: *for any k, $1 \leq k < n$, k is in the core of some microrun.*
Then by Lemma 1, x has a micro-R-cover $\{r_i : 1 \leq i \leq m\}$. If the position 1 is not in the core of at least two microruns, then $\mu(x[1..n]) \leq 1 + \mu(x[2..n])$ and so by the induction hypothesis $\mu(x[1..n]) \leq 1 + \mu(x[2..n]) \leq 1 + (n-1) = n$.
So we can assume that position 1 is in the core of at least two microruns. It follows that r_1 must be a microsquare with no tail and that there is a microsquare s with a period $<$ the period of r_1, starting at position 1, and no tail. Thus the conditions of Lemma 2 are fulfilled and so there is a cut k. It follows that $\mu(x) \leq k + \mu(s[k+1..n) = k + (n-k) = n$. ■

4 Conclusion and further research

We presented an alternative computational proof of the estimate of the number of microruns. The method presented does not scale up well for higher periods – though Lemma 2 holds as is for periods ≤ 10 – for higher periods more than just two initial squares of the R-cover are needed before the cut is guaranteed.

However, the most interesting aspect of R-covers was not fully exploited here: for a given string $x[1..n]$, if there is a position k in x that is not in the core of at least two microruns, than $\mu(x[1..n]) \leq \mu(s[1..k-1]) + 1 + \mu(x[k+1..n])$ and so by the induction hypothesis $\mu(x) \leq n$. This indicates that induction breaks down only if a string has two micro-R-covers, one a refinement of the other. There is a hope (and computational results carried to date provide some evidence), that such double covers are not possible. The future research will thus focus on a non-computational proof that such double covers do not exist providing a route to a non-computational estimate of the number of microruns.

BIBLIOGRAPHY

[1] M. CROCHEMORE: *An optimal algorithm for computing the repetitions in a word.* Inform. Process. Lett., 5 (5) 1981, pp. 297–315.

[2] M. CROCHEMORE AND L. ILIE: *Maximal repetitions in strings.* to appear in J. Comput. Syst. Sci.

[3] FAN KANGMIN AND W. F. SMYTH: *A new periodicity lemma.* to appear in SIAM J. of Discr. Math.

[4] F. FRANEK, J. SIMPSON, AND W. F. SMYTH: *The maximum number of runs in a string* in Proceedings of 14th Australasian Workshop on Combinatorial Algorithms AWOCA 2003, Seoul National University, Seoul, Korea, July 13-16 2003.

[5] F. FRANEK AND Q. YANG: *An asymptotic lower bound for the maximal number of runs in a string* Int. Journ. of Foundations of Computer Science, 1 (19) 2008, pp. 195–203.

[6] M. GIRAUD: *Not so many runs in strings* The proceedings of the LATA 2008, Tarragona, Spain, March 2008.

[7] C. S. ILIOPOULOS, D. MOORE, AND W. F. SMYTH: *A characterization of the squares in a Fibonacci string* Theoretical Computer Science 172 (1997) 281-291.

[8] R. KOLPAKOV AND G. KUCHEROV: *On maximal repetitions in words.* J. of Discrete Algorithms, (1) 2000, pp. 159–186.

[9] R. K. KUSANO, W. MATSUBARA, A. ISHIMO, H. BANNAI, AND A. SHINOHARA *New lower bounds for the maximum number of runs in a string*, CoRR, abs/0804.1214, May 2008, http://arxiv.org/abs/0804.1214,

[10] M. G. MAIN: *Detecting leftmost maximal periodicities.* Discrete Applied Maths., (25) 1989, pp. 145–153.

[11] S. J. PUGLISI, W. F. SMYTH, AND A. TURPIN: *Some restrictions on periodicity in strings*, in Proceedings of 16th Australasian Workshop on Combinatorial Algorithms AWOCA 2005, University of Ballarat, Victoria, Australia, September 18-21 2005, pp. 263–268.

[12] W. RYTTER: *The number of runs in a string: Improved analysis of the linear upper bound*, in Proceedings of 23rd Annual Symposium on Theoretical Aspects of Computer Science STACS 2006, Marseille, France, February 23-25 2006, pp. 184–195.

Frantisek Franek
Department of Computing & Software
McMaster University
Hamilton, Ontario, Canada L8S 4K1
Email: franek@mcmaster.ca
http://www.cas.mcmaster.ca/~franek

Jan Holub
Department of Computer Science and Engineering
Faculty of Electrical Engineering
Czech Technical University in Prague
Karlovo Namesti 13, 121 35 Prague 2, Czech Republic
Email: holub@fel.cvut.cz
http://cs.felk.cvut.cz/~holub

A Word Counting Graph

MIREILLE RÉGNIER, ZARA KIRAKOSYAN, EVGENIA FURLETOVA,
AND MIKHAIL ROYTBERG

ABSTRACT.   We study methods for counting occurrences of words
from a given set $\mathcal{H}$ over an alphabet V in a given text. All words have
the same length m. Our goal is the computation of the probability to
find p occurrences of words from a set $\mathcal{H}$ in a random text of size n,
assuming that the text is generated by a Bernoulli or Markov model.
We have designed an algorithm solving the problem; the algorithm
relies on traversals of a graph, whose set of vertices is associated with
the overlaps of words from $\mathcal{H}$. Edges define two oriented subgraphs
that can be interpreted as equivalence relations on words of $\mathcal{H}$. Let
$\mathcal{P}(\mathcal{H})$ be the set of equivalence classes and S be the set of other
vertices. The run time for the Bernoulli model is $O(np(|\mathcal{P}(\mathcal{H})|+|S|))$
time and the space complexity is $O(pm|S| + |\mathcal{P}(\mathcal{H})|)$. In a Markov
model of order K, additional space complexity is $O(pm|V|^{K})$ and
additional time complexity is $O(npm|V|^{K})$. Our preprocessing uses
a variant of the Aho-Corasick automaton and achieves $O(m|\mathcal{H}|)$ time
complexity. Our algorithm is implemented and provides a significant
space improvement in practice. We compare its complexity to the
additional improvement due to Aho-Corasick minimization.

1   Introduction

Studies on word probabilities started as early as the eighties with the seed
paper [12]. A recent interest arose from applications to computational biology [11, 14, 10, 19, 28]. Numerous statistical softwares [4, 25, 2] have
been designed recently to extract "In Silico" exceptional words , i.e. words
that are either overrepresented or underrepresented in genomes. An international competition was organized by M. Tompa [26] to evaluate their
capabilities and weaknesses. All softwares combine an algorithm to extract
candidate motifs and a statistical criterium to evaluate overrepresented or
underrepresented motifs. Sensitivity and selectivity of such criteria turn out
to be crucial, as well as the speed and easiness of computation. A survey
on motif searching can be found in [8].

We address here the following problem, that is fundamental to assess
significance. Given the alphabet V and a set $\mathcal{H}$ of words on alphabet V,

all words in $\mathcal{H}$ have the same length m. Our aim is to compute the probability to find exactly p occurrences of a word from a set $\mathcal{H}$ in a text of size n. Below, $t_n^{[p]}(\mathcal{H})$ denotes this probability. One assumes the text is randomly generated according to a Bernoulli or Markov model of order K. Naive solutions to this problem when $p = 1$ [5] lead to a $O(n|V|^K|\mathcal{H}|^2)$ time complexity. Recent states of the art of various improvements can be found in [24, 15]. On the one hand, the language approach defined in [21] allows for an elimination of $|V|^K$ multiplicative factor in [20]. On the other hand, several approximations can be derived for $t_n^{[1]}(\mathcal{H})$. They imply, at some stage, the computation of all possible overlaps of two words of $\mathcal{H}$ or a formal inversion of a $|\mathcal{H}| \times |\mathcal{H}|$ matrix of polynoms. Therefore, time complexity is $O(|\mathcal{H}|^2)$. A recent algorithm [13] allows for the computation of the generating function $\sum_{n \geq 0} t_n^{[1]}(\mathcal{H})z^n$ in $O(nm|\mathcal{H}|)$ time and $O(m|\mathcal{H}|)$ space complexity. Interestingly, $O(nm|\mathcal{H}|)$ time complexity outperforms $O(|\mathcal{H}|^2)$ complexity for most practical values of n. Indeed, sets may be rather big [27], especially when they are defined through a Position Specific Scoring Matrix or PSSM [23]. Other algorithms simulate an automaton. This automaton is derived from a transition study [7] or language equations in [16]; it is embedded in a Markov chain, using sparse matrices in [9, 17]. Algorithm AHOPRO [3] addresses a more general problem. It computes the probability to find several occurrences (up to fixed bound p_i for set $\mathcal{H}_i$) in several sets $\mathcal{H}_1, \cdots, \mathcal{H}_q$. It simulates an Aho-Corasick like automaton. Its time and space complexity are $O(n(|S| + |\mathcal{H}|)|V| \prod p_i)$ and $O((|S| + |\mathcal{H}|) \times \prod p_i)$, where $|S|$ is the size of the automaton. A rough upper bound for this size is $(m - 1)|\mathcal{H}|$. This reduces to $O(pn(|S| + |\mathcal{H}|))$ and $O(p(|S| + |\mathcal{H}|))$ for p occurrences in a single set. A recent publication [22] simulates the same automaton and implements a $\log n$ time complexity improvement, for a single set.

One common feature of these approaches is their easy extension to the Markov model. The multiplicative factor $|V|^K$ in time complexity [5] becomes an *additive* factor, except for [16]. The main difference comes from the *space* complexity. As mentioned in [22], Markov chain embeddings may be costly, due to matrices sparsity. Although automaton implementations are less sparse, space complexity remains an actual drawback [3].

One possible way to reduce space requirement is the classical minimization of the underlying automaton. This is realized, for p occurrences in a single set $\mathcal{H}$, in [18], with an $O(np|S||V|)$ time complexity. In this paper, we propose an alternative modelling of word counting based on graphs that are related to overlaps of words of the set. We define a partition of set $\mathcal{H}$ into equivalence overlap classes that we represent in graphs and we show that the set of equivalence overlap classes, denoted $\mathcal{P}(\mathcal{H})$, is efficiently computed in a preprocessing step from a variant of classic Aho-Corasick tree, that can be built in $O(m|\mathcal{H}|)$ time. Let S denote the set of internal nodes, that are associated to proper overlaps. We also show that probabilities $t_n^{[p]}(\mathcal{H})$ satisfy induction equations that only depend on these overlap classes. A simple algorithm follows, that relies on classical tree traversals with simple additional data structures that optimize memory management. As a whole, this algorithm improves on [16, 7, 13, 17, 3] as it achieves a $O(np(|\mathcal{P}(\mathcal{H})| + |S|))$ time computation and $O(pm|S|)$ space complexity with a *smaller* set S. Here, S is the set of vertices of the overlap graph that are not in $\mathcal{P}(\mathcal{H})$. Finally, this algorithm can be extended to address the same features as AHOPRO. This drastic space improvement on space and time complexity is discussed in 5. We also compare it with the alternative minimization approach [18]. Finally, we suggest on one example a possible combination with Aho-Corasick minimization.

2   Overlap Graphs

We will use terms *word* and *text* as synonyms; informally, a text is long, and a word is short. A pattern $\mathcal{H}$ of length m is a set $\mathcal{H} = \{H_1, \cdots, H_q\}$ of words such that all words have the same length m.

Given a word w, let $|w|$ denote its length. Given two words w and t, one notes $w \prec t$ iff w is a proper prefix of t and $w \subset t$ iff w is a proper suffix of t.

DEFINITION 1. Given a pattern $\mathcal{H}$ over an alphabet V, a word w is a *suffixprefix* word for $\mathcal{H}$ **iff** exists H, F in $\mathcal{H}$ such as

$$w \subset \mathrm{H} \text{ and } w \prec \mathrm{F} . \tag{1}$$

The set of suffixprefix words of a set $\mathcal{H}$ is called its *overlap set* and denoted $OV(\mathcal{H})$.

In all examples, we use DNA alphabet $V = \{A, C, G, T\}$.

EXAMPLE 1. Let $\mathcal{H}$ be the set

$$\mathcal{H} = \{ \quad H_1 = ACATATA, H_2 = AGACACA, H_3 = ATACACA,$$
$$H_4 = ATAGATA, H_5 = CATTATA, H_6 = CTTTCAC,$$
$$H_7 = CTTTCCA, H_8 = TACCACA \} .$$

Overlap set is $OV(\mathcal{H}) = \{ATA, ACA, TA, CA, A, AC, C, \epsilon\}$. One has, for example, $ACA \subset H_2$ and $ACA \prec H_1$.

One observes that prefix relation and suffix relation define partial orders on $OV(\mathcal{H}) \cup \mathcal{H}$. Clearly, the empty sequence ϵ is a prefix (respectively a suffix) of any suffixprefix and any word from $\mathcal{H}$. Moreover, the set of prefixes (respectively suffixes) of a word H in $\mathcal{H}$ is totally ordered. Thus, it admits a maximal element in $OV(\mathcal{H})$. Therefore, these relations naturally define equivalence relations between words in set $\mathcal{H}$ and, consequently, a partition of set $\mathcal{H}$.

DEFINITION 2. Given a pattern $\mathcal{H}$ over an alphabet V, the *left predecessor* of a word H, noted lpred(H), is its longest prefix belonging to the overlap set $OV(\mathcal{H})$. In other words,

$$\text{lpred(H)} = \max\{w \prec \text{H}, w \in OV(\mathcal{H})\} \ . \tag{2}$$

Analogously, the *right predecessor* of H, noted rpred(H), is its longest suffix belonging to the overlap set $OV(\mathcal{H})$. In other words,

$$\text{rpred(H)} = \max\{w \subset \text{H}, w \in OV(\mathcal{H})\} \ . \tag{3}$$

Two words H and F are said *left* (respectively *right*) *equivalent* iff they have the same left (respectively right) predecessor. For any pattern H, its left class is denoted $\bar{\text{H}}$ and its right class is denoted $\tilde{\text{H}}$.

Two words H and F are said *overlap equivalent* if they are both left and right equivalent. The *overlap class* (i.e. the class of overlap equivalence) of a word H is denoted $\dot{\text{H}}$. The set of all overlap classes is denoted $\mathcal{P}(\mathcal{H})$. One denotes lpred($\dot{\text{H}}$) (respectively rpred($\dot{\text{H}}$)) the common left (respectively right) predecessor of the patterns in $\dot{\text{H}}$.

EXAMPLE 2. In Example 1, $\tilde{\text{H}}_2 = \tilde{\text{H}}_3 = \tilde{\text{H}}_8$ as ACA is their largest *suffix* in $OV(\mathcal{H})$. There are four right classes $\tilde{C}_{ACA} = \{\text{H}_2, \text{H}_3, \text{H}_8\}$, $\tilde{C}_{ATA} = \{\text{H}_1, \text{H}_4, \text{H}_5\}$, $\tilde{C}_{AC} = \{\text{H}_6\}$ and $\tilde{C}_{CA} = \{\text{H}_7\}$.

$\bar{\text{H}}_3 = \bar{\text{H}}_4$ as ATA is their largest *prefix* in $OV(\mathcal{H})$. There are six left classes $\bar{C}_{ACA} = \{\text{H}_1\}$, $\bar{C}_{ATA} = \{\text{H}_3, \text{H}_4\}$, $\bar{C}_{CA} = \{\text{H}_5\}$, $\bar{C}_{TA} = \{\text{H}_8\}$, $\bar{C}_C = \{\text{H}_6, \text{H}_7\}$, $\bar{C}_A = \{\text{H}_2\}$.

There are eight overlap classes in $\mathcal{P}(\mathcal{H})$, that are the eight singletons of $\mathcal{H}$. In other words, $\mathcal{P}(\mathcal{H}) = \mathcal{H}$ and the partition is trivial.

DEFINITION 3. The *left overlap graph* is the oriented graph $LOG_{\mathcal{H}}$ built on the set of vertices $OV(\mathcal{H}) \cup \mathcal{P}(\mathcal{H})$ with an edge from vertex s to vertex t iff:

$$s = \text{lpred}(t) = \max\{w \prec t, w \in OV(\mathcal{H})\} \ . \tag{4}$$

The *right overlap graph* is the oriented graph $ROG_{\mathcal{H}}$ built on the set of vertices $OV(\mathcal{H}) \cup \mathcal{P}(\mathcal{H})$ with an edge from vertex s to vertex t iff

$$s = \text{rpred}(t) = \max\{w \subset t, w \in OV(\mathcal{H})\} \ . \tag{5}$$

The *overlap graph* $GOV_{\mathcal{H}}$ is the oriented graph $LOG_{\mathcal{H}} \cup ROG_{\mathcal{H}}$.

As all words in an overlap class have the same predecessor in the prefix and suffix relations, this definition makes sense for vertices t that represent overlap classes. Moreover, one observes that all predecessors of a word are totally ordered and that the empty string is a minimal element for this order. Therefore, both left overlap graph and right overlap graphs are *trees* where the root is associated to the empty string ϵ that has no predecessor. The leaves are the elements of $\mathcal{P}(\mathcal{H})$. This definition implies the property below.

PROPERTY 4. Given a word H in $\mathcal{H}$, or a class $\dot{\mathrm{H}}$ in $\mathcal{P}(\mathcal{H})$, its ancestors in the *left* (respectively *right*) *overlap graph* are its prefixes (respectively suffixes) that belong to $OV(\mathcal{H})$.

EXAMPLE 3. In Example 1, ancestors of H_1 and H_6 in left overlap graph, depicted in Figure 1, distinct from the root are $\{ACA, AC, A\}$ and $\{C\}$ respectively. In the right overlap graph, depicted in Figure 2, their ancestors distinct from the root, are $\{ATA, TA, A\}$ and $\{AC, C\}$. The paths associated to a node in $LOG_{\mathcal{H}}$ (respectively $ROG_{\mathcal{H}}$) are read from the top to the node (respectively, from the node to the top).

REMARK 1. Let us call *deep nodes* are the predecessors of overlap classes. In automaton theory, deep nodes represent left and right quotients of the recognized language, e.g. $\mathcal{H}$.

One observes that AC represents a right class and does not represent a left class. Nevertheless, the largest words in $OV(\mathcal{H})$, e.g. ACA and ATA, represent a right class and a left class.

EXAMPLE 4. Let $\mathcal{G}$ be $\{ACNNNTA\}$ where N stands for any symbol in alphabet V. Pattern $\mathcal{G}$ contains $4^3 = 64$ words and its overlap set contains 6 words: $OV(\mathcal{G}) = \{A, ACTA, ACATA, ACCTA, ACGTA, ACTTA, \epsilon\}$. There are 12 overlap classes in $\mathcal{P}(\mathcal{G})$. For each class, the left predecessor is underlined and the right predecessor is given in bold. Nine overlap classes reduce to one element: $\dot{G}_3 = \{\underline{AC}\mathbf{ACATA}\}$, $\dot{G}_4 = \{\underline{AC}\mathbf{ACCTA}\}$, $\dot{G}_5 = \{\underline{AC}\mathbf{ACGTA}\}$, $\dot{G}_6 = \{\underline{AC}\mathbf{ACTTA}\}$, $\dot{G}_7 = \{\underline{ACT}\mathbf{ACTA}\}$, $\dot{G}_9 = \{\underline{ACAT}\mathbf{ATA}\}$, $\dot{G}_{10} = \{\underline{ACCT}\mathbf{ATA}\}$, $\dot{G}_{11} = \{\underline{ACGT}\mathbf{ATA}\}$, and $\dot{G}_{12} = \{\underline{ACTT}\mathbf{ATA}\}$. Three overlap classes contain several words: $\dot{G}_2 = \{\underline{AC}[ACG]\mathbf{ACTA}\}$ and $\dot{G}_8 = \{\underline{ACTA}[AGT]\mathbf{TA}\}$. The remaining 49 words of the pattern $\mathcal{G}$ form the last overlap class $\dot{G}_1$: left and right predecessors of the class are equal to A. Interestingly, the minimal automaton that recognizes $\mathcal{G}$ has a larger number of states, namely 19.

Below, we define unions of overlap classes, as they appear in equations of words occurrences probabilities.

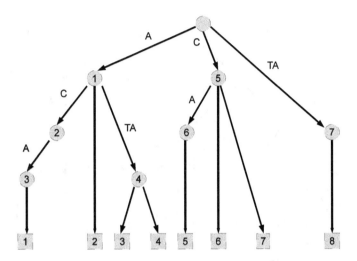

Figure 1. Left overlap graphs for Example 1. Round nodes are the elements of $OV(\mathcal{H})$ and leaves are the elements of $\mathcal{P}(\mathcal{H})$. In that particular case, since $\mathcal{P}(\mathcal{H}) = \mathcal{H}$, each leaf i is even simply the word H_i.

DEFINITION 5. Given a set $\mathcal{H}$ over an alphabet V and a word w in $OV(\mathcal{H})$, let $\bar{T}_w$ and $\tilde{T}_w$ be the subtrees rooted in w in the left and right overlap graphs, respectively. One defines the subsets of $\mathcal{P}(\mathcal{H})$:

$$\tilde{C}_w = \tilde{T}_w \cap \mathcal{P}(\mathcal{H}) , \qquad (6)$$
$$\bar{C}_w = \bar{T}_w \cap \mathcal{P}(\mathcal{H}) . \qquad (7)$$

A set $\tilde{C}_w$ is called a *right union* and a set $\bar{C}_w$ is called a *left union*.

REMARK 2. Given a right union $\tilde{C}_w$ and an overlap class $\dot{H}$ in $\mathcal{P}(\mathcal{H})$, the set $\dot{H} \cap \tilde{C}_w$ is either empty, or equal to $\dot{H}$ when w is a suffix of any word in $\dot{H}$. When w is a maximal suffixprefix, $\bar{C}_w$ and $\tilde{C}_w$ are overlap classes.

It follows from these definitions that a right (respectively left) union contains the leaves of tree $\tilde{T}_w$ (respectively $\bar{T}_w$). Therefore, Equations (8) and (9) follow immediately.

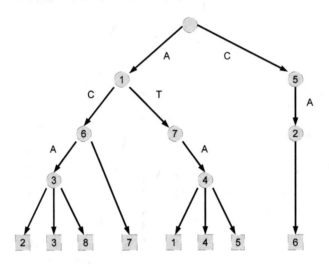

Figure 2. Right overlap graphs for Example 1.

$$\tilde{C}_w = \cup_{x \in ROG_{\mathcal{H}}, w \subseteq x} \tilde{C}_x = \cup_{w \subset H}(\dot{H}) \ , \qquad (8)$$
$$\bar{C}_w = \cup_{x \in LOG_{\mathcal{H}}, w \preceq x} \bar{C}_s = \cup_{w \prec H}(\dot{H}) \ . \qquad (9)$$

In Example 1, $\tilde{C}_A = \tilde{C}_{ACA} \cup \tilde{C}_{ATA} \cup \tilde{C}_{CA} = \{H_1, H_2, H_3, H_4, H_5, H_7, H_8\}$ and $\bar{C}_A = \bar{C}_{ACA} \cup \bar{C}_{ATA} \cup \bar{C}_A = \{H_1, H_2, H_3, H_4\}$. In Example 4, one has $\tilde{C}_{ACTA} = \dot{G}_2 \cup \dot{G}_7$.

Our bottom-up computations in 4.1 and 4.2 rely on these union properties.

REMARK 3. The relation $\sim$ defined as $x \sim y$ iff $xt \in \mathcal{H}$ is equivalent to $yt \in \mathcal{H}$ is well-known in automata theory as right $\mathcal{H}$-equivalence. There is a bijection between the states of the minimal automaton that recognizes $\mathcal{H}$ and classes of right equivalence.

3 Equations for words occurrences probabilities

The aim of this section is to establish our master Theorems 14 and 24 which state induction relations for word occurrence probabilities, in the Bernoulli and Markov model. They rely on a factorization lemma to achieve announced time and space complexities to be discussed in 5.1.

3.1 Basic Notations and Equations

One assumes the text **T** may be generated according to either one of the two models below.

(B) BERNOULLI MODEL: every symbol s of a finite alphabet V is created independently of the other symbols, with probability p_s.

(M) MARKOV MODEL: in a Markov model of order K, the probability of a symbol occurrence at some position in the text depends on the K previous symbols.

In a Markov model of order K, let $P_w(H)$ denote the probability to find H right of a given word w of length K. A Bernoulli model is viewed as a Markov model where K is 0. Given a word H, one denotes $P(H)$ its probability in the Bernoulli model. Given a set of words $\mathcal{F}$, one denotes $P(\mathcal{F}) = \sum_{H \in \mathcal{F}} P(H)$ and $P_w(\mathcal{F}) = \sum_{H \in \mathcal{F}} P_w(H)$ its probability in the Bernoulli and the Markov models, respectively.

DEFINITION 6. Let $\mathcal{H} = \{H_1, H_2, \ldots, H_q\}$ be a given set of words with the same length m over an alphabet V.

One denotes $t_n^{[p]}(\mathcal{H})$ (respectively $f_n^{[p]}(\mathcal{H})$) the probability to find exactly (respectively at least) p occurrences of $\mathcal{H}$ words in a text of size n.

One denotes $r_n^{[p]}(H)$ the probability to find word H in text **T**, with its end at position n, under the condition that exactly $(p-1)$ other patterns from $\mathcal{H}$ appeared among positions $1, \cdots, n-1$.

For a subset $\mathcal{F}$ of $\mathcal{H}$, one denotes $r_n^{[p]}(\mathcal{F})$ the probability to find a word of $\mathcal{F}$ ending at position n in text **T** with exactly $(p-1)$ $\mathcal{F}$-words before it, i.e. $r_n^{[p]}(\mathcal{F}) = \sum_{H \in \mathcal{F}} r_n^{[p]}(H)$.

PROPERTY 7. For any integers n and p, one has

$$t_n^{[p]}(\mathcal{H}) = f_n^{[p]}(\mathcal{H}) - f_n^{[p+1]}(\mathcal{H}) ,$$

$$f_n^{[p]}(\mathcal{H}) = \sum_{k=0}^{n} r_k^{[p]}(\mathcal{H})$$

Therefore, the computation of $t_n^{[i]}$ follows from the computation of $(r_n^{[i]}(H))$ where H ranges over $\mathcal{H}$.

DEFINITION 8. Given a word t and a prefix w of t, one denotes $\Phi_w(t)$ the suffix of t that satisfies $t = w\Phi_w(t)$. One denotes $\phi_w(t)$ the probability of $\Phi_w(t)$ in the Bernoulli model. By convention, $\phi_w(t) = 0$ if w is not a prefix of t.

Given a set $\mathcal{F}$ of words with a left common ancestor w, one defines

$$\phi_w(\mathcal{F}) = \sum_{F \in \mathcal{F}} \phi_w(F) \ . \tag{10}$$

EXAMPLE 5. For word H_1 from set $\mathcal{H}$ in Example 1, one has $\Phi_{ACA}(H_1) = TATA$; then, in the Bernoulli model, $\phi_{ACA}(H_1) = p_T^2 p_A^2$.

Proposition 9 below rewrites, using unions in overlap graphs defined above, a classical equation for words in $\mathcal{H}$ derived for $p = 1$ in [24, 15, 13] for the Bernoulli model. This rewriting will incorporate classes of overlap equivalence in Theorems 14 and allows for a further generalization to Markov models in 24. These two theorems allow for an implementation based on overlap graphs with announced complexities.

PROPOSITION 9. *Assume a Bernoulli model. Let $\mathcal{H}$ be a given set of words and p be an integer. The probabilities $(r_n^{[1]}(H))_{n \geq m}$ where H ranges in $\mathcal{H}$ satisfy the induction relation:*

$$r_n^{[1]}(H) = \left[1 - \sum_{k=1}^{n-m} r_k^{[1]}(\mathcal{H}) \right] P(H) \tag{11}$$
$$- \sum_{w \in LOG_{\mathcal{H}} \backslash \{\epsilon\}, w \prec H} r_{n-m+|w|}^{[1]}(\tilde{C}_w)\phi_w(H) \ .$$

For any integer i satisfying $2 \leq i \leq p$, the probabilities $(r_n^{[i]}(H))_{n \geq m}$ where H ranges in $\mathcal{H}$ satisfy the induction relations:

$$r_n^{[i]}(H) = \sum_{k=1}^{n-m} \left[r_k^{[i-1]}(\mathcal{H}) - r_k^{[i]}(\mathcal{H}) \right] P(H) \tag{12}$$
$$+ \sum_{w \in LOG_{\mathcal{H}} \backslash \{\epsilon\}, w \prec H} \left[r_{n-m+|w|}^{[i-1]}(\tilde{C}_w) - r_{n-m+|w|}^{[i]}(\tilde{C}_w) \right] \phi_w(H) \ ,$$

with initial conditions:

$$r_1^{[i]}(H) = \cdots = r_{m-1}^{[i]}(H) = 0 \ , \quad 1 \leq i \leq p \ ,$$
$$r_m^{[1]}(H) = P(H) \ ,$$
$$r_m^{[i]}(H) = 0 \ , \quad 2 \leq i \leq p \ .$$

Proof. We propose a proof of this Proposition in the Bernoulli model, when $p = 1$, that extends to p occurrences, $p \geq 1$. This Proposition is extended to overlap classes and to Markov model in Theorems 14 and 24. We study events that lead to a first occurrence and establish their probabilities. A word H is the first occurrence of $\mathcal{H}$ in text $\mathbf{T}$ at position n iff

(i) word H occurs, ending at position n;

(ii) no $\mathcal{H}$ word occurred up to position $n - m$;

(iii) no $\mathcal{H}$ word occurred in positions $n - m + 1, \cdots, n - 1$.

The probability of (ii) is $1 - \sum_{k=1}^{n-m} r_k^{[1]}(\mathcal{H})$. Term $(1 - \sum_{k=1}^{n-m} r_k^{[1]}(\mathcal{H}))P(\mathrm{H})$ is the probability of the simultaneous occurrence of (i) and (ii), as these events are independent in the Bernoulli model. One must substract the probability of the complementary event of (iii), conditioned by (i) and (ii). Complementary event of (iii) conditioned by (ii) is the union of $(m - 1)$ disjoint events, $(E_l)_{1 \leq l \leq m-1}$, where each event E_l represents a first occurrence of a $\mathcal{H}$-word ending at position $n - m + l$. Each event E_l contains (i). Now, event E_l conditioned by (i) occurs iff a word F of $\mathcal{H}$ occurs at position $n - m + l$ when H ends at position n. This is satisfied iff exists w, with $|w| = l$ and $l \neq 0$, that is a prefix of H and a suffix of F. Equivalently, w is a left ancestor of H different from the root and F belongs to $\tilde{C}_w$.

This reasoning steadily extends. To find an i-th occurrence ending at position n, $(i - 1)$ occurrences, are needed before $n - m$; an i-th occurrence is forbidden before position $n - m$ (condition (ii)), and between $n - m + 1$ and $n - 1$ (condition (iii)). ∎

EXAMPLE 6. Equation for word $\mathrm{H}_1 \in \mathcal{H}$ is:

$$r_n^{[1]}(\mathrm{H}_1) = [1 - \sum_{k=1}^{n-7} r_k^{[1]}(\mathcal{H}))]P(\mathrm{H}_1) - r_{n-6}^{[1]}(\tilde{C}_A)P(CATATA)$$
$$- r_{n-5}^{[1]}(\tilde{C}_{AC})P(ATATA) - r_{n-4}^{[1]}(\tilde{C}_{ACA})P(TATA) .$$

As each word has no more than m ancestors, the recursive equations (11) and (12) provide an easy way to compute $(r_n^{[p]}(\mathrm{H}))_{\mathrm{H} \in \mathcal{H}}$ in $O(npm|\mathcal{H}|)$ time complexity. This complexity is achieved in [13] with a matrix and in [3] with an Aho-Corasick-like automaton. In [18], the automaton is minimized, which yields a complexity $O(np|V||S|)$, where the size of the automaton is smaller. Overlap graphs provide an alternative improvement.

3.2   Overlap Graph Traversal

Our time complexity improvement, i.e. $O(np(|\mathcal{P}(\mathcal{H})| + |S|))$, where $S = OV(\mathcal{H})$, relies on a factorization property introduced below that reduces the computation to two graph traversals in Lemmas 11, 12 and 21. Space complexity issues will be discussed in Section 4 below.

Graph traversal Equations (11) and (12) are the same for all words H in $\mathcal{H}$, up to the coefficients P(H) and ϕ_w(H). Lemma 11 establishes that the computation of the second term in (11) or(12) for each *word* H or *left class* $\bar{\mathrm{H}}$ only requires *one* parameter $(\psi_n^{[p]}(t))$ defined for its left predecessor t in $LOG_{\mathcal{H}}$ and that this parameter can be computed in a traversal of the graph.

DEFINITION   10.   In   the   Bernoulli   model,   one   defines $\{(\psi_n^{[i]}(w))_{w \in OV(\mathcal{H}) \setminus \{\epsilon\}}\}_{1 \le i \le p}$ by the p top-down inductions

$$\psi_n^{[i]}(w) = \begin{cases} r_{n-m+|w|}^{[i]}(\tilde{C}_w), & \text{if lpred}(w) = \epsilon \\ \psi_n^{[i]}(\text{lpred}(w)) \cdot \phi_{\text{lpred}(w)}(w) & \text{otherwise .} \\ \quad + r_{n-m+|w|}^{[i]}(\tilde{C}_w), \end{cases} \tag{13}$$

Lemma 11 establishes that $\psi_n^{[i]}(w)$ represents the information on the past that is shared by all descendants of w.

LEMMA 11. *For any word* H *in* $\mathcal{H}$, *and any left ancestor* w *of* H *such that* $w \ne \epsilon$, *one has:*

$$\sum_{x \in OV(\mathcal{H}) \setminus \{\epsilon\}, x \preceq w} r_{n-m+|x|}^{[i]}(\tilde{C}_x)\phi_x(\mathrm{H}) = \psi_n^{[i]}(w) \cdot \phi_w(\mathrm{H}) \ . \tag{14}$$

Any subset $\mathcal{T}$ *of a left class* $\bar{\mathrm{H}}$ *in* $\mathcal{H}$ *satisfies*

$$\sum_{\mathrm{F} \in \mathcal{T}} \sum_{w \in OV(\mathcal{H}) \setminus \{\epsilon\}, w \prec \mathrm{F}} r_{n-m+|w|}^{[i]}(\tilde{C}_w)\phi_w(\mathrm{F}) = \psi_n^{[i]}(t) \cdot \phi_t(\mathcal{T}) \ , \tag{15}$$

where t *is the common left predecessor.*

Proof. Indeed, ϕ_x(H) rewrites $\phi_x(w)\phi_w$(H). Therefore, $\psi_n^{[i]}$ inductive definition allow to prove inductively, for any i in $[1..p]$, that

$$\sum_{w \in OV(\mathcal{H}) \setminus \{\epsilon\}, w \prec \mathrm{H}} r_{n-m+|w|}^{[i]}(\tilde{C}_w)\phi_w(\mathrm{H}) = \psi_n^{[i]}(t) \cdot \phi_t(\mathrm{H}) \ , \tag{16}$$

where t is the left predecessor of H in $LOG_{\mathcal{H}}$. A summation over all words in a subset T of left class $\bar{\mathrm{H}}$ and a factorization yields (15). ∎

Second term in (11) or (12) may be computed once for any subset of words in a left class. Yet, probabilities computation for larger n implies to memorize separately right classes. Therefore, we split left classes with respect to right equivalence, which yields classes of $\mathcal{P}(\mathcal{H})$. A rewriting of Equation (8) leads to Lemma 12 below.

LEMMA 12. *Let w be a node in $ROG_\mathcal{H}$. For any integer i or n, probabilities $r_n^{[i]}(\tilde{C}_w)$ satisfy:*

$$r_n^{[i]}(\tilde{C}_w) = \sum_{x \in OV(\mathcal{H}), w \subseteq x} r_n^{[i]}(\tilde{C}_x) = \sum_{\dot{H} \subseteq \tilde{C}_w} r_n^{[i]}(\dot{H}) . \tag{17}$$

COROLLARY 13. *The set $\{r_n^{[i]}(\tilde{C}_w)\}_{w \in OV(\mathcal{H})}$ can be computed from $\{r_n^{[i]}(\tilde{C}_x)\}_{x \in D_r}$ by a bottom-up traversal of $ROG_\mathcal{H}$.*

THEOREM 14. *Let $\dot{H}$ be a class in $\mathcal{P}(\mathcal{H})$ and denote t the left predecessor of $\dot{H}$. In the Bernoulli model, probabilities $r_n^{[i]}(\dot{H})$ satisfy*

$$r_n^{[1]}(\dot{H}) = \left[1 - \sum_{k=1}^{n-m} r_k^{[1]}(\mathcal{H})\right] P(\dot{H}) - \psi_n^{[1]}(t) \cdot \phi_t(\dot{H}), \tag{18}$$

and, for $i \geq 2$,

$$r_n^{[i]}(\dot{H}) = \sum_{k=1}^{n-m} \left[r_k^{[i-1]}(\mathcal{H}) - r_k^{[i]}(\mathcal{H})\right] P(\dot{H}) + \left[\psi_n^{[i-1]}(t) - \psi_n^{[i]}(t)\right] \cdot \phi_t(\dot{H}), \tag{19}$$

with initial conditions

$$\begin{aligned} r_1^{[i]}(H) = \cdots = r_{m-1}^{[i]}(H) &= 0 , \quad 1 \leq i \leq p , \\ r_m^{[1]}(H) &= P(H) , \\ r_m^{[i]}(H) &= 0 , \quad 2 \leq i \leq p . \end{aligned}$$

3.3 Markov extension

Unlike the algorithms based on finite automata accepting the corresponding texts (see e.g. [16, 7, 13, 18, 3, 22]), our computation on the Markov model is a generalization of (11) that relies onto overlap graphs introduced above. An extension of the induction to the Markov model requires to extend the $(\psi_n^{[i]})$ family for nodes w with a depth smaller than K. This is realized with a suitable *partition* of union $\tilde{C}_w$. We use the notations of Definition 8 and restrict the presentation to the case $K \leq m$.

DEFINITION 15. Given a word w such that $K \leq |w|$, one denotes $suf_K(w)$ its suffix of size K. Given a word α in V^K, one denotes

$$\mathcal{H}(\alpha) = \{H \in \mathcal{H}; \alpha = suf_K(H)\} \ .$$

Given a word t such that w is a prefix of t, one defines parameter $\phi_w(t)$ by

$$\phi_w(t) = P_{suf_K(w)}(\Phi_w(t)) \ . \tag{20}$$

REMARK 4. $\phi_w(t)$ represents the probability to find $\Phi_w(t)$ right after a w-occurrence. In Example 1, $\phi_{ACA}(H_1) = P_{CA}(T)P_{AT}(A)P_{TA}(T)P_{AT}(A)$ in the Markov model of order 2.

DEFINITION 16. Given a set $\mathcal{H}$ and a Markov model of order K, with $K \leq m$, one defines the *right K-frontier* of $\mathcal{H}$, noted $RF_K(\mathcal{H})$, and the *left K-frontier*, noted $LF_K(\mathcal{H})$, by:

$$\begin{aligned} RF_K(\mathcal{H}) &= \{w \in OV(\mathcal{H}); |\mathrm{rpred}(w)| < K \leq |w|\} \ , &(21)\\ LF_K(\mathcal{H}) &= \{w \in OV(\mathcal{H}); |\mathrm{lpred}(w)| < K \leq |w|\} \ . &(22) \end{aligned}$$

The set of *right K-leaves*, noted $RL_K(\mathcal{H})$, and the set of *left K-leaves*, noted $LL_K(\mathcal{H})$, are defined as:

$$\begin{aligned} RL_K(\mathcal{H}) &= \{\dot{H} \in \mathcal{P}(\mathcal{H}); |\mathrm{rpred}(\dot{H})| < K\} \ , &(23)\\ LL_K(\mathcal{H}) &= \{\dot{H} \in \mathcal{P}(\mathcal{H}); |\mathrm{lpred}(\dot{H})| < K\} \ . &(24) \end{aligned}$$

EXAMPLE 7. In Example 1, the right 3-frontier is $\{ACA, ATA\}$ and the right 3-leaves are $\dot{H}_6$ and $\dot{H}_7$.

The right K-frontier and the set of right K-leaves allow to define for every word w such that $|w| < K$, a natural partition of each union $\tilde{C}_w$ defined by the successors of w in $ROG_\mathcal{H}$ that belong to these two sets.

DEFINITION 17. Given a word w in $OV(\mathcal{H}) \setminus \{\epsilon\}$ such that $|w| < K$, one defines a subset $S_K(w)$ of V^K as

$$S_K(w) = \bigcup_{t \in RF_K(\mathcal{H}) \cup RL_K(\mathcal{H}), w \subset t} \{suf_K(t)\} \ . \tag{25}$$

Given a word w in $RF_K(\mathcal{H}) \cup RL_K(\mathcal{H})$, one denotes

$$S_K(w) = \{suf_K(w)\} \ .$$

EXAMPLE 8. Let w be A in Example 1. Word t will range in $\{ACA, ATA, H_7\}$. Therefore $S_3(A) = \{ACA, ATA, CCA\}$. Let w be AC. Then word t ranges in $\{H_6\}$ and $S_3(AC) = \{CAC\}$.

REMARK 5. With the notations above, one gets

$$\tilde{C}_w = (\cup_{\{t \in RF_K(\mathcal{H}); w \subseteq \mathrm{rpred}(t)\}} \tilde{C}_t) \bigcup (\cup_{\{\dot{H} \in RL_K(\mathcal{H}); w \subseteq \mathrm{rpred}(\dot{H})\}} (\dot{H})) \ .$$

For instance, it follows from Example 8 that $\tilde{C}_{CA} = \tilde{C}_{ACA} \cup \{H_7\}$.

DEFINITION 18. Given a word w in $OV(\mathcal{H}) \setminus \{\epsilon\}$ satisfying $|w| < K$, one defines for each α in $S_K(w)$:

$$r^{[i]}_{n,\alpha}(w) = \sum_{\{t \in RF_K(\mathcal{H}); w \subseteq \mathrm{rpred}(t), \alpha \subseteq t\}} r^{[i]}_n(\tilde{C}_t) \qquad (26)$$

$$+ \sum_{\{\dot{H} \subset RL_K(\mathcal{H}); w \subseteq \mathrm{rpred}(\dot{H}), \alpha \subseteq \dot{H}\}} r^{[i]}_n(\dot{H}) \ .$$

By convention, $r^{[i]}_{n,\alpha}(w) = 0$ if $\alpha \in V^K \setminus S_K(w)$.

REMARK 6. It is guaranteed that at least one of the two sums contribute, and that $w \subset \alpha$.

We now show how this knowledge on the past allows for the computation of Equations (11) and (12).

DEFINITION 19. Given a word w in $OV(\mathcal{H}) \setminus \{\epsilon\}$ satisfying $|w| < K$, one denotes $SP_K(w)$ the subset of V^K defined inductively as follows.

$$SP_K(w) = \begin{cases} S_K(w) & \text{if } \mathrm{lpred}(w) = \epsilon \\ S_K(w) \cup \{\alpha \in V^K; S^{[\alpha]}(w) \neq \emptyset\} & \text{otherwise ,} \end{cases} \qquad (27)$$

where

$$S^{[\alpha]}(w) = \{\beta \in SP_K(\mathrm{lpred}(w)); \alpha = suf_K(\beta \cdot \Phi_{\mathrm{lpred}(w)}(w))\} \ . \qquad (28)$$

EXAMPLE 9. In Example 1, one has $SP_3(A) = S_3(A) = \{ACA, ATA, CCA\}$. Let w be AC. One has $\mathrm{lpred}(AC) = A$, and α ranges over $\{CAC, TAC\}$ when β ranges over $SP_3(A)$. Finally, $S^{[CAC]}(AC) = \{ACA, CCA\}$ and $S^{[TAC]}(AC) = \{ATA\}$ that are a partition of $SP_3(A)$. Finally, $SP_3(AC) = \{CAC, TAC\}$.

DEFINITION 20. Let w be any word in $OV(\mathcal{H}) \setminus \{\epsilon\}$.

If $|w| < K$, one defines a family $(\psi^{[p]}_{n,\alpha}(w))_{\alpha \in SP_K(w)}$.

$$\psi^{[p]}_{n,\alpha}(w) = \begin{cases} r^{[p]}_{n-m+|w|,\alpha}(w) & \text{if } \mathrm{lpred}(w) = \epsilon \\ \sum_{\beta \in S^{[\alpha]}(w)} \psi^{[p]}_{n,\beta}(x) P_\beta(\Phi_x(w)) & \text{otherwise .} \\ \quad + r^{[p]}_{n-m+|w|,\alpha}(w) \end{cases} \qquad (29)$$

If $|w| \geq K$, one defines a single function $\psi_n^{[p]}(w)$.

$$
\psi_n^{[p]}(w) \;=\; r_{n-m+|w|}^{[p]}(\tilde{C}_w) \tag{30}
$$
$$
+ \begin{cases} \sum_{\beta \in S^{[suf_K(w)]}(w)} \psi_{n,\beta}^{[p]}(x) P_\beta(\Phi_x(w)) & \text{if } w \in LF_K(\mathcal{H}) \\ \psi_n^{[p]}(x) \cdot \phi_x(w) & \text{otherwise .} \end{cases}
$$

EXAMPLE 10. Let us use (26), (29) and (30) in Example 1.

  (i) Let w be A. As $\mathrm{lpred}(A) = \epsilon$, one applies the first equation in (29) for α ranging in $S_3(A)$. Right members are computed through (26) which yields $\psi_{n,ACA}^{[p]}(A) = r_{n-6}^{[p]}(H_2) + r_{n-6}^{[p]}(H_3) + r_{n-6}^{[p]}(H_8)$, $\psi_{n,ATA}^{[p]}(A) = r_{n-6}^{[p]}(H_1) + r_{n-6}^{[p]}(H_4) + r_{n-6}^{[p]}(H_5)$ and $\psi_{n,CCA}^{[p]}(A) = r_{n-6}^{[p]}(H_7)$.

 (ii) One uses (29) to compute $\psi_{n,TAC}^{[i]}(AC)$ and $\psi_{n,CAC}^{[i]}(AC)$. For instance,

$$
\psi_{n,CAC}^{[i]}(AC) \;=\; \psi_{n,ACA}^{[i]}(A) \cdot P_{ACA}(C) + \psi_{n,CCA}^{[i]}(A) \cdot P_{CCA}(C)
$$
$$
+ r_{n-5,CAC}^{[i]}(AC) \;,
$$

   where, using (26), $r_{n-5,CAC}^{[i]}(AC) = r_{n-5}^{[i]}(\dot{H}_6)$.

(iii) Word A is the left predecessor of ATA that belongs to the right 3-frontier. Therefore, (30) yields

$$
\psi_n^{[p]}(ATA) \;=\; \psi_{n,ACA}^{[p]}(A) P_{ACA}(TA) + \psi_{n,ATA}^{[p]}(A) P_{ATA}(TA)
$$
$$
+ \psi_{n,CCA}^{[p]}(A) P_{CCA}(TA) + r_{n-4}^{[p]}(H_1) + r_{n-4}^{[p]}(H_4)
$$
$$
+ r_{n-4}^{[p]}(H_5) \;.
$$

Lemma 11 can now be rewritten into Lemma 21 below that steadily yields Theorem 24.

LEMMA 21. *Any leaf class $\dot{H}$ in $\mathcal{P}(\mathcal{H})$ with left predecessor t satisfies*

$$
\sum_{w \in LOG_{\mathcal{H}} \backslash \{\epsilon\}, w \prec H} r_{n-m+|w|}^{[p]}(\tilde{C}_w) \phi_w(\dot{H}) \;=\; \tag{31}
$$
$$
\begin{cases} \psi_n^{[p]}(t) \cdot \phi_t(\dot{H}) & \text{if } |t| \geq K \;, \\ \sum_{\alpha \in V^K, t \subset \alpha} \psi_{n,\alpha}^{[p]}(t) \cdot \phi_t(\dot{H}) & \text{if } |t| < K \;. \end{cases}
$$

DEFINITION 22. Let $f_k^{[p]}(\alpha)$ be the probability that a text of size k, $k \geq m$, ending with suffix α in V^K, contains at least p occurrences of words from $\mathcal{H}$.

LEMMA 23. The probabilities $f_k^{[p]}(\alpha)$ satisfy the following induction

$$f_k^{[p]}(\alpha) = \sum_{\beta \in V^K} f_{k-1}^{[p]}(\beta) \cdot P_\beta(\alpha) + r_k^{[p]}(\mathcal{H}(\alpha)) \ , \tag{32}$$

with the initial condition

$$f_m^{[p]}(\alpha) = r_m^{[p]}(\mathcal{H}(\alpha)) \ .$$

THEOREM 24. Let $\dot{H}$ be a class in $\mathcal{P}(\mathcal{H})$ and t be its left predecessor. In the Markov model, probabilities $r_n^{[i]}(\dot{H}(\beta))$ are equal to

$$\sum_{\alpha \in V^K} (f_{n-m}^{[i-1]}(\alpha) - f_{n-m}^{[i]}(\alpha)) P_\alpha(\dot{H}(\beta)) \qquad + \tag{33}$$

$$\begin{cases} (\psi_n^{[i-1]}(t) - \psi_n^{[i]}(t)) \cdot \phi_t(\dot{H}) & if \quad |t| \geq K \\ \sum_{\alpha \in V^K} (\psi_{n,\alpha}^{[i-1]}(t) - \psi_{n,\alpha}^{[i]}(t)) \cdot \phi_t(\dot{H}) & if \quad \dot{H} \subset LL_K(\mathcal{H}) \end{cases}$$

with initial conditions

$$r_1^{[i]}(H) = \cdots = r_{m-1}^{[i]}(H) \ = \ 0 \ , \quad 1 \leq i \leq p \ ,$$
$$r_m^{[1]}(H) \ = \ P(H) \ ,$$
$$r_m^{[i]}(H) \ = \ 0 \ , \quad 2 \leq i \leq p \ ,$$

and convention $\psi_n^{[0]}(t) = 0$ and $\psi_{n,\alpha}^{[0]}(t) = 0$.

4   Basic Algorithm

We describe a procedure to compute the probabilities of words occurrences in texts. Our input is a set of words $\mathcal{H}$. Our output is the set of probabilities $(r_n^{[i]}(\mathcal{H}))_{1 \leq i \leq p}$ for a given position n. Overlap graph modelling provides a straighforward but efficient algorithm.

4.1   Overlap Graphs Construction

Our preprocessing step builds right and left overlap graphs and partition $\mathcal{P}(\mathcal{H})$. It relies on a tree $\mathcal{T}_\mathcal{H}$ built with Aho-Corasick algorithm [1]. A basic feature of Aho-Corasick is the on-line construction of so-called *suffix links*. Given a word s in V^*, suf(s) denotes the largest suffix of s that is a proper prefix of some word from the set $\mathcal{H}$. Below, we identify each node or leaf with the word defined by the path from the root. A suffix link is a link from a node s to node suf(s).

Right overlap graph We start with the right overlap graph. According to Definition 1, a node in $\mathcal{T}_{\mathcal{H}}$ belongs to $OV(\mathcal{H})$ iff there exists a sequence of suffix links from a leaf to this node. Therefore, a bottom-up traversal of $\mathcal{T}_{\mathcal{H}}$ according to the suffix links allows one to find out the nodes from $OV(\mathcal{H})$ and to construct $ROG_{\mathcal{H}}$. See Figure 3.

Left overlap graph To link the nodes in $OV(\mathcal{H})$ with the links for $LOG_{\mathcal{H}}$, one can use a depth-first traversal of $\mathcal{T}_{\mathcal{H}}$.

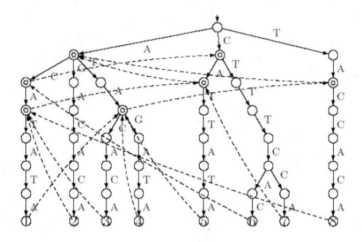

Figure 3. Aho-Corasick tree $\mathcal{T}_{\mathcal{H}}$ for set $\mathcal{H}$ in Example 1 used to build $ROG_{\mathcal{H}}$ and $LOG_{\mathcal{H}}$ in Figures 1 and 2. The nodes corresponding to $OV(\mathcal{H})$ are marked with double circles. A leaf corresponding to the word H_i in $\mathcal{H}$ is labeled with the number $i, 1 \leq i \leq 8$. The dashed lines depicts suffix links needed to construct $ROG_{\mathcal{H}}$; the other Aho-Corasicks suffix links are omitted.

Partition Thanks to the order of AC tree, a depth-first traversal of $LOG_{\mathcal{H}}$, *MaxSufPref*, achieves a partition of $\mathcal{H}$ into $\mathcal{P}(\mathcal{H})$. As all words of a class are left equivalent, it is enough to identify below each deep node x all words that are right equivalent and number the designed classes. To achieve this linearly, one maintains in each internal node s the class number i(s) of the last visited leaf H in $\mathcal{H}$ satisfying suf(H) = s. One achieves an increasing numbering; let j denote the number of the last class created before the visit of leaves below x. When a leaf H is visited, an equivalent leaf has been previously visited iff $i(\mathrm{suf}(H)) > j$. In this case, leaf H is addressed to class $i(\mathrm{suf}(H))$. Otherwise, it is given the next available class number k and $i(\mathrm{suf}(H))$ is updated to k.

Markov model Main preprocessing step is the computation of the right K-frontier and the right K-leaves, which is $O(|V|^K)$.

Aho-Corasick construction of the tree, including the suffix links, can be achieved in time and space $O(m|\mathcal{H}|)$ [6]. Subgraph extraction algorithms are classical graph traversals. They run with $O(|\mathcal{T}_\mathcal{H}|)$ time and space complexity and $O(|\mathcal{T}_\mathcal{H}|)$ is upper bounded by $O(m|\mathcal{H}|)$ [1]. Once these steps have been performed, tree $\mathcal{T}_\mathcal{H}$ is not needed any more and space can be released.

4.2   Algorithm and basic data structures

Our algorithm computes inductions (19) and (34) in a depth-first traversal of $LOG_\mathcal{H}$. Bottom-up traversals of $ROG_\mathcal{H}$ allow to manage memory: an extraction of suitable information is performed at the beginning of each cycle and an update is realized at the end of each cycle. For sake of clarity, we first describe the Bernoulli model.

Memory management One observes that results of previous computations that are used in (19) are $\sum_{k=1}^{n-m}(r_k^{[i-1]}(\mathcal{H}) - r_k^{[i]}(\mathcal{H}))$ and $(\psi_n^{[i]}(t))$ that depend on $(r_{n-m+|w|}^{[i]}(\tilde{C}_w))_{w \in LOG_\mathcal{H} \setminus \{\epsilon\}}$.

1. Values $\sum_{k=1}^{n-m} r_k^{[i]}(\mathcal{H})$ where i ranges in $\{1 \cdots p\}$ are memorized in p global variables, e.g. an array $SumProb$ of p integers.

2. For each node w at depth $|w|$ in $LOG_\mathcal{H} \setminus \{\epsilon\}$, one needs values $(r_{n-m+|w|}^{[i]}(\tilde{C}_w))_{1 \le i \le p}$ to perform computation at cycle n. Therefore, one memorizes $p(m-|w|)$ values $\{r_{n-m+|w|}^{[i]}(\tilde{C}_w), \cdots, r_{n-1}^{[i]}(\tilde{C}_w)\}_{1 \le i \le p}$ in an array $ProbMark$ of size $p(m-|w|)$.

3. In each node w and in each leaf class $\dot{H}$ in $LOG_\mathcal{H} \setminus \{\epsilon\}$, one memorizes probability $\phi_x(w)$ or $\phi_x(\dot{H})$, where x is its father in $LOG_\mathcal{H}$.

Algorithm For any inductive step n, our algorithm executes:

1. In a bottom-up traversal of $ROG_\mathcal{H}$, including the root, $ProbMainCalc$ extracts from $ProbMark$, for each internal node, the p values $(r_{n-m+|w|}^{[i]}(\tilde{C}_w))_{1 \le i \le p}$. They are addressed in a field of each node. At the root, $(r_n^{[i]}(\mathcal{H}))$ are extracted.

2. In a depth-first traversal of $LOG_\mathcal{H}$:

   (a) when one visits an internal node, $(\psi_n^{[i]}(w))_{1 \le i \le p}$ are computed according to (13);

   (b) when a leaf class $\dot{H}$ is visited:

(i) compute $r_n^{[i]}(\dot{H})$ using (19), for i in $\{1, \cdots, p\}$;

(ii) Add $r_n^{[i]}(\dot{H})$ to a temporary variable, $FirstTemp[i]$, in its predecessor in $ROG_{\mathcal{H}}$ (a deep node).

3. In a bottom-up traversal of $ROG_{\mathcal{H}}$, update $ProbMark$ for internal nodes, using $FirstTemp$ according to (17). At the root ϵ, $r_n^{[i]}(C_\epsilon) = r_n^{[i]}(\mathcal{H})$ is added to $SumProb$.

Markov model For memory management, it is now necessary to store values $\sum r_k^{[i]}(\mathcal{H}(\beta))$ for all words β in V^K. Moreover, for a node w at depth smaller than K, one stores, for each i, non-zero values $r_{n,\alpha}^{[i]}(w)$. For a node w in the left K-frontier or above it, and for a leaf class $\dot{H}$ in the left K-leaves, one must memorize several values $P_\beta(\Phi_x(w))$ or $P_\beta(\Phi_t(\dot{H}))$.

Procedures are slightly modified. In nodes above the right K-frontier, one extracts additional values, namely $r_{n,\alpha}^{[i]}(w)$, in step 1. In step 2, families $(\psi_n^{[i]}(w))$ computation relies on (29) and necessitates a preprocessing to memorize all $\phi_x(w)$.

If a leaf class $\dot{H}$ is a left K-leaf, one computes $r_n^{[i]}(\dot{H})$ through (34) instead of (19). In step 3, the bottom-up traversal stops at the right K-frontier.

5   Complexity and experimental results

5.1   Complexity

Overlap graph complexity Our algorithm achieves an overall $O(np(|S|+|\mathcal{P}(\mathcal{H})|))$ time complexity, where $S = OV(\mathcal{H})$. Indeed, depth-first traversal at step 2 and bottom-up traversals at steps 1 and 3 yield a $O(p|S|)$ time complexity. Computation of (19) is done for each overlap class in constant time. This yields an $O(|\mathcal{P}(\mathcal{H})|)$ time complexity in the Bernoulli model.

Temporary memory for $\psi_n^{[p]}(w)$ computation or probability update is $O(p|S|)$. In our current implementation, $ProbMark$ is built for all internal nodes, which yields a $O(mp|S|)$ space complexity. This already improves on recent algorithms [13, 3]. This upper bound can be improved. Indeed, given a node w at depth k, one needs only memorize $(r_{n-l}^{[i]}(\tilde{C}_w))$ when for $1 \le i \le p$ and $1 \le l \le m - k$. Therefore, space complexity is pD with

$$D = \sum_{w \in S} [m - |w|] \ . \tag{34}$$

When a Markov model is used, additional space is needed. First, m values $f_k^{[i]}(\alpha)$ use $O(pm|V|^K)$ space. Then, for a node w, constraint $w \subset \alpha$ yields at most $|V|^{K-|w|}$ non-zero values $r_{n,\alpha}^{[i]}$. Additional space is $O(pK|V|^K)$.

Finally, computation of (29) needs values $P_\beta(\Phi_x(w))$. A node x needs at most $|V|^{K-|x|}$ values $P_\beta(\Phi_x(w))$. Summing over at most $|V|^{|x|}$ nodes at level $|x|$ yields an upper bound $O(pK|V|^K)$. Therefore, overall additional space complexity is $O(pm|V|^K)$. As each term is computed once at each inductive step, in $O(1)$ time, additional time complexity is $O(npm|V|^K)$.

Previous algorithms Exact computation of p occurrences probability depends on the text size n as a linear function $O(n)$ or a logarithmic function $O(\log n)$. Let us mention that there also exist *approximate* computations that may be realized in constant time with some drawbacks on the space complexity or the tightness of the approximation. They are beyond the scope of this paper.

We now discuss exact computation. All approaches, including automata or Markov Chain embedding [16, 7, 9, 18], matrices [13] or languages [20] need to compute a set of *linear* equations of order m with *constant* coefficients. Therefore, theoretical complexity is known. Algorithms that are linear in n, such as REGEXPCOUNT, AHOPRO or SPATT achieve a complexity $p|S|n$ where $|S|$ is the size of the structure used for the *computation*. Using the classical system reduction [20], it is theoretically possible to achieve a $O(m^p \log n)$ algorithm, with a constant factor that depends of the data structure to be used. This optimization has been recently implemented in [22]. Nevertheless, one observes that the m^p factor may represent a significant drawback. Therefore, counting algorithms are compared below on the basis of the two data structures used in the computation and the memorization.

[13] data structure is $O(|\mathcal{H}| \log |\mathcal{H}|)$. Markov Chain Embedding algorithms use very sparse matrices. Therefore, REGEXPCOUNT and AHOPRO automata outperform them, and the minimization achieved in [18] is a further improvement. Overlap graph is an alternative approach. In both cases, there is no non-trivial upper bound for the ratio between $|S|$ and the size of Aho-Corasick automata. Remark that one must take into account the alphabet size V in automata time complexity and a multiplicative factor m in overlap graph space complexity. When a minimization is very efficient, (pattern $ANNNNA$), an overlap graph is less significantly reduced. Our last example suggests that building an overlap graph on a minimized automaton is an improvement and might be optimal. Indeed, for set $\mathcal{G}$ in Example 4, minimized Aho-Corasick automaton has 3 final states and 19 internal nodes. Overlap graph for this minimized automaton contains 4 classes and 4 internal nodes, associated to prefixes $\epsilon, C, CC = GG$ and $CCCC$. This difference is mainly due to backward links, that are necessary for a searching algorithm, but are unnecessary for counting algorithms, and therefore erased by overlap graphs.

5.2   Experiments

We have performed computer experiments to compare our approach and the approach using finite automaton representation of the corresponding set of text (e.g. [3, 18]). We used a Intel Celeron 2.26 GHz with Windows XP. The complexity of automaton based methods can be described with the size $|\Sigma|$ of minimal automaton accepting the set of texts containing the desired number of patterns. The proposed approach leads to other characteristics of the pattern:

(i) $|S|$ is the number of overlaps;

(ii) $|\mathcal{P}(\mathcal{H})|$ is the number of *overlap classes*, i.e. equivalence classes in the overlap-based partition of the given word set;

(iii) D is the sum of differences $m - |w|$ over all suffix-prefixes w defined in (34).

In our case, it was stated above in section 5.1) that the run-time is proportional to $|S| + |\mathcal{P}(\mathcal{H})|$ and the required space is proportional to D.

First, we studied the dependence of the size $|S|$ of overlap graph on the size and length of the pattern $\mathcal{H}$. Data are given in Table 2 and depicted in Figure 4 Surprisingly, the average value of $|S|$ for random patterns does not depend on the pattern length m while the size of the minimal automaton grows almost linearly with the growth of the pattern length (the number of words in the pattern being fixed). One can see that for reasonable pattern lengths (up to 24) and sizes (up to 1500 words) the value $|S|$ (and even $|S| + |\mathcal{P}(\mathcal{H})|$) is less than the number $|\Sigma|$ of states of the corresponding minimized automaton. The independence of value $|S|$ of the pattern length can be justified theoretically (see Theorem 25 in Section 6).

Figure 5 demonstrates the impact of usage of overlap classes instead of individual words (cf. equations (11) and (19)). We have generated 30 sets of $50k$ random words ($k = 1, 2, \cdots, 30$) of length 8 and 12; the random words were generated in 4-letter alphabet $\{A, C, G, T\}$ using Bernoulli model with equal probabilities of all letters. In the vast majority of cases, the number of overlap classes coincides with the initial number of words. We give the line only for the random patterns of length 8; the line for random patterns of length 12 coincides with it. Two lower lines corresponds to the patterns based on two position-specific scoring matrices (PSSM). Given a PSSM $\mathcal{M}$, a threshold or cutoff s defines a set $\mathcal{F}(\mathcal{M}, s)$, that consists of all the words in V^m with a score greater than s. Different cutoffs were used to obtain patterns of different sizes. The PSSMs were created by Dan Pollard for Drosophila genes (http://www.danielpollard.com/matrices.html) and are

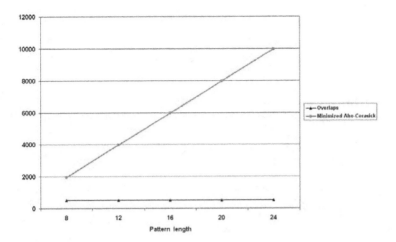

Figure 4. Overlap graph size $|S|$ and minimal automaton size $|\Sigma|$. Average values for random patterns of size 500. Blue diamond shapes represent $|S|$ and purple squares represent $|\Sigma|$.

given in Table 1. The PSSMs are referred to as PSSM-08 and PSSM-12; their lengths are 8 and 12, respectively. Cutoffs are chosen to ensure that the set sizes range from 1 to 1500. One can see that in case of non-random patterns usage of overlap classes may bring significant improvement.

Tables 3, 4, 5 and 6 demonstrate results of experiments with the above four series of patterns. For each pattern we show its length and size, as well as above complexity measures related both to automaton-based and overlap-based approaches. Columns *Time* and *Space* represent time and space complexity of algorithms. For the automaton-based approach the run-time is proportional to $n \cdot p \cdot |V| \cdot |\Sigma|$ and the required space is proportional to $p \cdot |\Sigma|$ (see [18]); here n is the length of the text and p is the desired number of occurrences. As we have shown in section 5.1, in our approach we get analogous formulas, namely $O(n \cdot p \cdot (|S| + |\mathcal{P}(\mathcal{H})|))$ for time and $O(p \cdot D)$ for space. Thus, in our comparison we use $|V| \cdot |\Sigma|$ and $|\Sigma|$ as time and space coefficients for automaton-based algorithm; we use $|S| + |\mathcal{P}(\mathcal{H})|$ and D as analogous coefficients for the overlap-based approach.

In Figures 6, 7, 8 and 9, one can see that in the vast majority of cases the coefficients for overlap approach are less than ones for the automaton approach. Time coefficients and space coefficients for PSSM-08 patterns

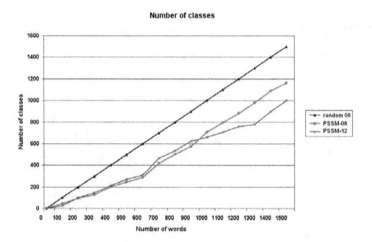

Figure 5. Number of overlap classes as a function of the number of words in the pattern. The data are given for three series of patterns: (1) random patterns of length 8; (2) patterns corresponding to two PSSMs with different cutoffs. The upper line (blue triangles) corresponds to random patterns.

are depicted in Figures 6 and 7, respectively.   The average ratios of time

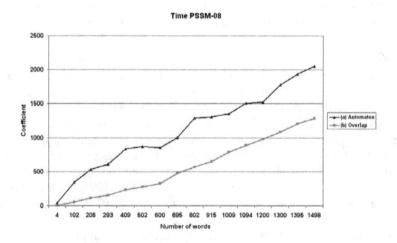

Figure 6. Time coefficients of the patterns determined by PSSM-08.

-	A	C	G	T
1	0.405	-1.424	0.301	-0.097
2	-0.944	0.122	-0.097	0.454
3	-0.622	-0.622	-3.989	1.067
4	1.041	-3.989	-1.156	-0.182
5	1.353	-3.989	-3.989	-2.380
6	-3.989	-1.792	-1.792	1.294
7	-3.989	-0.770	0.054	0.901
8	0.702	-3.989	0.665	-3.989

-	A	C	G	T
1	0.368	-2.197	-0.588	0.636
2	0.000	0.000	0.000	0.000
3	0.636	-2.197	-2.197	0.636
4	-2.197	-2.197	-2.197	1.299
5	1.299	-2.197	-2.197	-2.197
6	-2.197	-2.197	-2.197	1.299
7	-2.197	1.299	-2.197	-2.197
8	-2.197	-2.197	1.299	-2.197
9	1.022	0.000	-2.197	-2.197
10	0.000	-0.588	-2.197	0.847
11	0.847	-0.588	-0.588	-0.588
12	0.636	-0.588	0.368	-2.197

Table 1. Position-specific scoring matrices (PSSM) built to study Drosophila genes (http://www.danielpollard.com/matrices.html) of lengths 8 and 12.

coefficients are 4.64 (for patterns of length less than 500); 2.12 (for patterns with lengths between 500 and 1000); 1.65 (for patterns of length between 1000 and 15004). The plot for PSSM-12 is similar (the plot is not shown, see Tables 3 and 4). The advantage of overlap approach for space coefficients (in case of PSSM-08 patterns) is not so high compared to time coefficients, but still significant (see Figure 7). However for the PSSM-12 patterns, the space coefficients for the automaton approach are in average slightly better (the plot is not shown, see Tables 3 and 4).

The data on random patterns (see Figures 8 and 9) demonstrate that in this case the overlap-based approach works better. The lines depicting the behavior of the overlap approach for random patterns of different lengths almost coincide. This corresponds to the results given in Table 2 and Figure

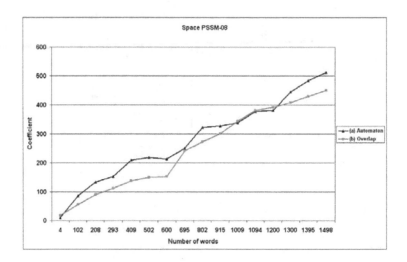

Figure 7. Space coefficients of the patterns determined by PSSM-08.

4. In contrast, the automaton approach coefficients for the patterns of length 12 are significantly greater than ones for the patterns of length 8.

6   Discussion

The majority of the algorithms computing probability to find the given number of occurrences of a pattern in a random text [16, 7, 9, 17, 3, 18] are based on the usage of the automaton accepting the appropriate set of texts. The structure of the automaton reflects the set of suffix-prefixes (overlaps) of the given pattern but relation between states of the minimal automaton and overlaps can be quite complicated.

In this paper we present another approach based on explicit usage of partition of the given word set according to the structure of its overlaps. Recursive equations, (11) used here and in [13], (12) and (19) employ subtraction of probabilities and therefore cannot be reduced to the automaton-based equations for any automaton.

The series of computer experiments allowed one to learn and compare the advantages of the approaches. For the overlap approach, the important characteristics of the pattern is number $|S|$ of its suffix-prefixes; the characteristics is also important for various pattern matching methods. The experiments demonstrate that for random patterns the value $|S|$ depend only of patterns size (i.e. number of words) but does not depend on pattern length (see Table 2 and Figure 4). The following Theorem justifies this observation.

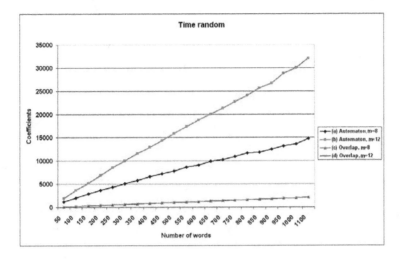

Figure 8. Time coefficients for the random patterns of lengths 8 and 12. The graphs for overlap coefficients of lengths 8 and 12 (the lowest lines) almost coincide.

THEOREM 25. *Let V be an alphabet and $U(r; m)$ be a set of all sets (patterns) of r words with the same length m over an alphabet V; all elements of $U(r; m)$ are considered as having the same probability. Let further $E(r, m)$ be the average number of suffix-prefixes (overlaps) of the patterns from $U(r; m)$. Then*

$$E(r, m) \leq C \cdot r \tag{35}$$

where C does not depend on r and m.

The proof of the theorem is not given here due to lack of space and will be presented in a companion paper.

The above feature predetermines that the proposed approach exceeds the automaton one in case of random patterns (Figures 8 and 9). For non-random patterns, the automaton approach due to general minimization procedure get benefits from the structure of pattern and in these cases the advantage of the overlap approach is not so clear; moreover, in some cases the automaton approach works better. It seems to be interesting to provide deeper investigation of the dependence of the methods behavior on the structure of patterns and to develop a new method combining the advantages of the approaches.

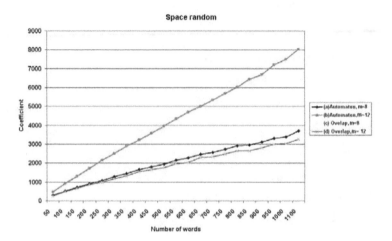

Figure 9. Space coefficients for the random patterns of lengths 8 and 12. The graphs for overlap coefficients of lengths 8 and 12 (the lowest lines) almost coincide.

7   Conclusion

In this paper, we introduced a new concept of overlap graphs to count word occurrences and their probabilities. The concept led to a recursive equation that differs from the classical one based on the finite automaton accepting the proper set of texts. In case of many occurrences, our approach yields the same order of time and space complexity as the approach based on minimized automaton.

For random patterns, our approach has asymptotically better constants and this advantage is achieved for relatively small pattern length, e.g. for lengths around 10 (see Figures 8 and 9). For non-random patterns, our approach gives results compatible with S-PATT [18]; the results depend on the structure of the pattern. Because of this, we here restricted ourselves with comparison of theoretical constants rather than run-times. The description of the algorithm in Sections 4 and 5 justifies this way of algorithms comparison. Remark that the constants well describe the space complexity of the algorithms. In the same time, our results are significantly better than the results of methods that do not implement the automaton minimization, e.g. [16, 7, 13, 17, 3].

In the future, we plan to design a novel recursion equation combining advantages of overlap graphs and minimal automata. Other directions of improvement are extensions of the scheme allowing to count differently various patterns and implement probability distributions described with Hidden Markov Models.

Acknowledgments The authors are very grateful to anonymous referees whose remarks greatly helped to improve this paper.

E. Furletova and M. Roytberg were supported by grants 08-01-92496-NCNIL-a and 09-04-01053- from RFBR (Russia). M. Régnier and M. Roytberg were supported by INTAS grant 05-1000008-8028 and MIGEC-INRIA associate team.

BIBLIOGRAPHY

[1] A. Aho and M. Corasick. Efficient String Matching. *CACM*, 18(6):333–340, 1975.

[2] B. Berman, B. Pfeiffer, T. Laverty, S. Salzberg, G. Rubin, M. Eisen, and S. Celniker. Computational identification of developmental enhancers: conservation and function of transcription factor binding-site clusters in drosophila melanogaster and drosophila pseudoobscura. *Genome Biol.*, 5(9), 2004. R61.

[3] V. Boeva, J. Clément, M. Régnier, M. Roytberg, and V. Makeev. Exact p-value calculation for heterotypic clusters of regulatory motifs and its application in computational annotation of cis-regulatory modules. *Algorithms for molecular biology*, 2(13):25 pages, 2007.

[4] K. Cartharius, K. Frech, K. Grote, B. Klocke, M. Haltmeier, A. Klingenhoff, M. Frisch, M. Bayerlein, and T. Werner. MatInspector and beyond: promoter analysis based on transcription factor binding sites. *Bioinformatics*, pages 2933–2942, 2005.

[5] C. Chrysaphinou and S. Papastavridis. The occurrence of sequence of patterns in repeated dependent experiments. *Theory of Probability and Applications*, 79:167–173, 1990.

[6] M. Crochemore and W. Rytter. *Jewels in Stringology*. World Scientific Publishing, Hong Kong, 2002.

[7] M. Crochemore and V. Stefanov. Waiting time and complexity for matching patterns with automata. *Inf. Process. Lett.*, 87(3):119–125, 2003.

[8] M. Das and H.-K. Dai. A survey of DNA motif finding algorithms. *BMC Bioinformatics*, 8(52):247, 2007.

[9] J. Fu and W. Lou. *Distribution theory of runs and patterns and its applications. A finite Markov chain imbedding approach*. World Scientific, Singapore, 2003. 162p., ISBN 981-02-4587-4.

[10] M. Gelfand and E. Koonin. Avoidance of Palindromic Words in Bacterial and Archaeal Genomes: a Close Connection with Restriction Enzymess. *Nucleic Acids Research*, 25(12):2430–2439, 1997.

[11] M. Gelfand, E. Koonin, and A. Mironov. Prediction of Transcription Regulatory Sites in Archaea by a Comparative Genome Approach. *Nucleic Acids Research*, 28:695–705, 2000.

[12] L. Guibas and A. Odlyzko. String Overlaps, Pattern Matching and Nontransitive Games. *Journal of Combinatorial Theory*, Series A, 30:183–208, 1981.

[13] L. Hertzberg, O. Zuk, G. Getz, and E. Domany. Finding motifs in promoter regions. *Journal of Computational Biology*, 12(3):314–330, 2005.

[14] D. S. Latchman. *Eukaryotic Transcription Factors*. Elsevier Academic Press, New York, 2004. 4-th edition, 360 pp., ISBN 0-12-437178-7.

[15] Lothaire. *Applied Combinatorics on Words*. Cambridge University Press, Reading, Mass., 2005.

[16] P. Nicodème, B. Salvy, and P. Flajolet. Motif statistics. *Theoretical Computer Science*, 287(2):593–618, 2002. preliminary version at ESA'99.

m	100		200		500		1000	
	$|\Sigma|$	$|S|$	$|\Sigma|$	$|S|$	$|\Sigma|$	$|S|$	$|\Sigma|$	$|S|$
8	510	96	929	199	1960	515	3427	988
12	906	98	1713	192	3976	518	7358	1006
16	1302	107	2533	207	5956	503	11368	961
20	1698	95	3307	190	7958	516	15415	987
24	2105	92	4114	201	9983	512	19442	991

Table 2. Average number of overlaps and size of minimized Aho-Corasick automaton for random patterns of different lengths (from 8 to 24) and sizes (from 100 to 1000 words). The number of overlaps does not depend on the pattern's length while the size of the automaton grows linearly with the length of the pattern.

[17] G. Nuel. Effective p-value computations using finite markov chain imbedding (fmci): application to local score and to pattern statistics. *Algorithms for molecular biology*, 1(5):14 pages, 2006.

[18] G. Nuel. Pattern markov chains: optimal markov chain embedding through deterministic finite automata. *J. Appl. Prob.*, 45:226–243, 2008.

[19] E. Panina, A. Mironov, and M. Gelfand. Statistical analysis of complete bacterial genomes:Avoidance of palindromes and restriction-modification systems. *Mol. Biol.*, 34:215–221, 2000.

[20] M. Régnier. A Unified Approach to Word Occurrences Probabilities. *Discrete Applied Mathematics*, 104(1):259–280, 2000. Special issue on Computational Biology;preliminary version at RECOMB'98.

[21] M. Régnier and W. Szpankowski. On Pattern Frequency Occurrences in a Markovian Sequence. *Algorithmica*, 22(4):631–649, 1997. preliminary draft at ISIT'97.

[22] P. Ribeca and E. Raineri. Faster exact Markovian probability functions for motif occurrences: a DFA-only approach. *Bioinformatics*, 24(24):2839–2848, 2008.

[23] G. D. Stormo. DNA binding sites: representation and discovery. *Bioinformatics*, 16:16–23, 2000.

[24] W. Szpankowski. *Average Case Analysis of Algorithms on Sequences*. John Wiley and Sons, New York, 2001.

[25] G. Thijs, K. Marchal, M. Lescot, S. Rombauts, B. D. Moor, P. Rouze, and Y. Moreau. A Gibbs sampling method to detect overrepresented motifs in the upstream regions of coexpressed genes. *Journal of Computational Biology*, 9:447–464, 2002.

[26] M. Tompa, N. Li, T. Bailey, G. Church, B. De Moor, E. Eskin, A. Favorov, M. Frith, Y. Fu, J. Kent, V. Makeev, A. Mironov, W. Noble, G. Pavesi, G. Pesole, M. Régnier, N. Simonis, S. Sinha, G. Thijs, J. van Helden, M. Vandenbogaert, Z. Weng, C. Workman, C. Ye, and Z. Zhu. An assessment of computational tools for the discovery of transcription factor binding sites. *Nature Biotechnology*, 23(1):137 – 144, Jan. 2005.

[27] H. Touzet and J.-S. Varré. Efficient and accurate p-value computation for position weight matrices. *Algorithms for Molecular Biology*, 15(2), 2007. 12 pages.

[28] M. Vandenbogaert and V. Makeev. Analysis of Bacterial RM-systems through Genome-scale Analysis and Related Taxonomic Issues. *In Silico Biology*, 3:12, 2003. preliminary version at BGRS'02.

	Automaton approach			Overlap approach			
$|\mathcal{H}|$	N_{AC}	$|\Sigma|$	Time	$|\mathcal{P}(\mathcal{H})|$	$|S|$	Time	Space
4	19	10	40	4	4	8	18
102	305	86	344	42	11	53	55
208	550	134	536	94	19	113	90
293	733	152	608	128	24	152	112
409	971	209	836	204	31	235	138
502	1159	218	872	244	34	278	149
600	1351	213	852	289	36	325	153
695	1528	251	1004	419	55	474	241
802	1740	322	1288	501	65	566	273
915	1940	327	1308	576	73	649	301
1009	2130	339	1356	707	85	792	344
1094	2288	377	1508	793	94	887	381
1200	2499	381	1524	883	97	980	392
1300	2689	445	1780	978	106	1084	409
1395	2860	484	1936	1089	113	1202	430
1498	3039	513	2052	1164	122	1286	450

Table 3. Characteristics of patterns related to PSSM-08 containing different numbers $|\mathcal{H}|$ of words. For the automaton approach, N_{AC} is the size of Aho-Corasick automaton; Σ is the size of minimal automaton that represents the coefficient for space complexity; coefficient for time complexity $Time$ equals $|\Sigma| \times |V|$. For the overlap approach we give the number of overlap classes, $\mathcal{P}(\mathcal{H})$; the number of overlaps S; time coefficient that is equal to $|S|+|\mathcal{P}(\mathcal{H})|$ and space coefficient $Space$ from (34).

	Automaton approach			Overlap approach			
$|\mathcal{H}|$	N_{AC}	$|\Sigma|$	Time	$|\mathcal{P}(\mathcal{H})|$	$|S|$	Time	Space
48	243	57	228	14	6	20	18
104	416	103	412	24	8	32	55
160	512	108	432	74	14	88	90
232	640	119	476	125	16	141	112
296	736	119	476	144	16	160	138
352	840	129	516	198	20	218	149
400	928	131	524	213	21	234	153
464	1016	133	532	237	21	258	241
528	1120	145	580	304	23	327	273
592	1224	148	592	312	23	335	301
664	1469	207	828	446	44	490	344
728	1533	205	820	488	44	532	381
768	1589	201	804	496	45	541	382
816	1725	234	936	563	50	613	383
880	1861	246	984	611	50	661	384
952	1973	219	876	638	51	689	385
984	2005	214	856	662	51	713	386
1184	2997	344	1376	754	58	812	392
1240	3077	345	1380	762	58	820	409
1328	3237	404	1616	792	60	852	430
1584	4213	561	2244	1111	83	1194	450

Table 4. Characteristics of patterns related to PSSM-12 containing different numbers of words.

	Automaton approach			Overlap approach			
$|\mathcal{H}|$	N_{AC}	$|\Sigma|$	Time	$|\mathcal{P}(\mathcal{H})|$	$|S|$	Time	Space
50	299	284	1136	50	51	101	256
100	549	512	2048	100	108	208	501
150	783	729	2916	149	150	299	660
200	1003	919	3676	200	208	408	868
250	1198	1085	4340	250	243	493	985
300	1410	1290	5160	300	301	601	1175
350	1592	1448	5792	350	341	691	1319
400	1814	1658	6632	400	417	817	1561
450	1993	1808	7232	450	454	904	1648
500	2156	1941	7764	500	488	988	1766
550	2362	2161	8644	550	561	1111	1973
600	2528	2276	9104	600	584	1184	2034
650	2722	2469	9876	650	674	1324	2291
700	2855	2564	10256	700	697	1397	2346
750	3050	2738	10952	750	746	1496	2483
800	3217	2914	11656	800	815	1615	2656
850	3336	2955	11820	850	799	1649	2641
900	3519	3130	12520	900	879	1779	2807
950	3693	3303	13212	950	947	1897	3002
1000	3823	3398	13592	1000	976	1976	3044
1100	4156	3699	14796	1100	1059	2159	3262
1200	4452	3970	15880	1200	1195	2395	3564
1300	4750	4202	16808	1300	1231	2531	3693
1400	5063	4502	18008	1400	1388	2788	4027
1500	5356	4790	19160	1500	1509	3009	4284

Table 5. Characteristics of random patterns of length 8 containing different numbers of words.

	Automaton approach			Overlap approach			
$|\mathcal{H}|$	N_{AC}	$|\Sigma|$	Time	$|\mathcal{P}(\mathcal{H})|$	$|S|$	Time	Space
50	496	476	1904	50	51	101	256
100	959	921	3684	99	95	194	501
150	1367	1311	5244	150	153	303	660
200	1801	1722	6888	199	200	399	868
250	2224	2142	8568	250	265	515	985
300	2613	2494	9976	300	288	588	1175
350	3007	2888	11552	349	370	719	1319
400	3396	3232	12928	400	395	795	1561
450	3756	3579	14316	450	452	902	1648
500	4170	3959	15836	500	489	989	1766
550	4552	4335	17340	550	567	1117	1973
600	4938	4694	18776	600	609	1209	2034
650	5294	5013	20052	650	627	1277	2291
700	5650	5339	21356	699	677	1376	2346
750	5985	5686	22744	750	736	1486	2483
800	6368	6022	24088	800	803	1603	2656
850	6776	6431	25724	850	887	1737	2641
900	7061	6675	26700	900	908	1808	2807
950	7514	7194	28776	950	994	1944	3002
1000	7854	7495	29980	1000	1021	2021	3044
1100	8499	8019	32076	1099	1060	2159	3262

Table 6. Characteristics of random patterns of length 12 containing different numbers of words.

Mireille Régnier
INRIA
78153 Le Chesnay, France
and
Laboratoire J.-V.Poncelet (UMI 2615)
119002, Bolshoy Vlasyevskiy Pereulok 11, Moscow, Russia
Email: mireille.regnier@inria.fr

Zara Kirakosyan
Yerevan State University
377 200 Yerevan, Armenia
Email: zarulinka@yahoo.com

Evgenia Furletova
Institute of Mathematical Problems of Biology
142290, Institutskaya, 4, Pushchino, Russia
and
Pushchino State University
142290, Prospect Nauki,5, Pushchino, Russia
Email: janny51@rambler.ru

Mikhail Roytberg
Institute of Mathematical Problems of Biology
142290, Institutskaya, 4, Pushchino, Russia
and
Pushchino State University
142290, Prospect Nauki,5, Pushchino, Russia
and
Laboratoire J.-V.Poncelet (UMI 2615)
119002, Bolshoy Vlasyevskiy Pereulok 11, Moscow, Russia
Email: mroytberg@gmail.com

An Approach for Identifying Salient Repetition in Multidimensional Representations of Polyphonic Music

Jamie Forth and Geraint A. Wiggins

ABSTRACT.   SIATEC is an algorithm for discovering patterns in multidimensional datasets [15]. This algorithm has been shown to be particularly useful for analysing musical works. However, in raw form, the results generated by SIATEC are large and difficult to interpret. We propose an approach, based on the generation of set-covers, which aims to identify particularly salient patterns that may be of musicological interest. Our method is capable of identifying principal musical themes in Bach Two-Part Inventions, and is able to offer a human analyst interesting insight into the structure of a musical work.

1   Introduction

This paper attempts to identify the repetition of perceptually salient patterns in symbolically represented music. A geometrical approach is adopted in which pieces of music are represented as multidimensional datasets. Following the work of [15, 16] and [14], we have implemented SIATEC, a pattern induction algorithm, and have conducted a series of use-case studies in order to investigate the properties of the generated results in terms of musicological value and perceptual salience. SIATEC is known to discover many more patterns than are typically of interest in any music analysis context, and the results can be difficult to interpret [15, p. 340]. We propose a post-processing step, similar in character to the NP-hard minimum weighted set cover problem [9], in which various heuristics can be employed in order to optimise the results in terms of specific music-analytic objectives.

Viewing this problem in the light of theoretical computer science is beneficial on several counts. Firstly, the knowledge that an exact solution cannot be computed in polynomial time assures us that we need to seek approximate solutions. Secondly, a great deal of research has been conducted in trying to establish methods for deriving solutions within acceptable bounds of approximation. The standpoint of computer science informs understanding of the nature of this problem, and provides examples of rigorously tested methods that may be applicable to our case. The inherent level of ambiguity in set-covering problems accords with the common situation in musical analysis whereby different interpretations of a work may be considered equally valid and correct. In order for an analyst to reach any firm conclusion, compromises must be made, which are often informed by conventions (heuristics) of music theory.

An applied aim of this research is to develop tools suitable for various music-analytic tasks. Within the field of musicology such tools may assist conventional score analysis, and may prove particularly useful for larger-scale corpus analysis. The latter overlaps with interests of music information retrieval, where such techniques may be applied in order to extract commonly occurring patterns as the basis for classification. A composer may also be able to gain inspiration by analysing a work in progress and arriving at a fresh perspective. More novel applications may be found in music psychology or artificial intelligence, where large collections of music could be analysed in order to derive data for training or testing models of musical behaviour. The work discussed in this paper has been conducted within a larger project investigating musical creativity. This wider project requires a considerable body of musical knowledge with which to train artificial musical agents in an attempt to simulate aspects of creative musical behaviour. In order to justify claims of potential utility, tools for automated musical analysis must be adaptable to individual tasks, which may involve handling varying degrees of ambiguity, and necessarily must scale to real-world musical problems.

2   Previous work

The concept of a musical pattern entails repetition. The definition of SIATEC ensures the enumeration of *all* maximal repeated patterns [15, pp. 331–333]. A large number of these discovered patterns will usually prove to be of little interest from a musical or perceptual perspective, and this is one problem our heuristics must address. Yet a more complicated issue concerns the many *types* of salient repetitive patterns that may exist in a musical work. In other words, the kinds of patterns that are likely to be of interest, and the ways in which they are interesting, may vary considerably.

There is agreement amongst both musicologists and music psychologists as to the importance of repetition in music (see, for example, [12, 17, 10]). One cross-cultural study based on fifty musical works found that 94 percent of all musical passages longer than a few seconds in duration were repeated at some point in the work [8, pp. 228–9]. However, this result does not account for the role of repetition in music in its entirety, because repetition may exist in many forms beyond the exact repetition of musical events in sequence. For example, melodies may still be perceived as instances of the same basic melodic motif despite being transposed in pitch or transformed in time. Indeed, perceptual similarity may pertain for any individual listener under an arbitrary number of processes of elaboration and transformation. In the context of computational analysis, therefore, careful consideration must be given to the notion of pattern equality.

Much previous work in this area has concentrated on techniques for string matching. Despite successes within certain specialised tasks, notably concerning monophonic melodies, various limitations of string methods become apparent in the context of music [11].

An alternative approach to string matching exists in the form of geometrically-based algorithms. Within a geometrical framework, the individual note events of a piece of music correspond to single points in a multidimensional space.[1] A family of algorithms related to SIA (*Structure Induction Algorithm*) have been developed for pattern discovery and matching in multidimensional datasets [15, 19, 16, 14]. The initial development of these techniques was motivated to a large extent for application to music, but are equally applicable in other domains where objects may be adequately represented in a multidimensional space.

Following [15, p. 328], we define a datapoint as a k-tuple of real numbers, and a pattern P or dataset D as a finite set of k-dimensional datapoints. We reserve the term *dataset* to refer to a complete set of datapoints we wish to process, for example, a piece of music, while *pattern* refers to a subset of the dataset. A translator is a vector that maps from one instance of a pattern to another within a dataset. More precisely, a vector t is a *translator* for P in D if and only if the translation of P by t is also a subset of D.

The basic SIA algorithm computes all the maximal repeated patterns in a dataset [15, pp. 334–5]. The algorithm finds the largest non-empty set of translatable datapoints for every positive translation possible within the dataset. Hence, each pattern discovered by SIA is called a *maximal translatable pattern* (MTP). The worst case running time of SIA is $O(kn^2 log_2 n)$.

An important extension to SIA is SIATEC [15, pp. 335–8]. SIATEC underlies both the approach to pattern discovery adopted in the present paper, as well as the closely related COSIATEC algorithm, which will be discussed below. Like SIA, the SIATEC algorithm enumerates all the maximal translatable patterns in a dataset, but also groups them into equivalence classes. A *translational equivalence class* (TEC) is represented as an ordered pair

---

[1] A variation on this approach, based on sets of line segments in space, is discussed in [18]

Composition	Number of datapoints	Number of TECs
BWV 772	458	9035
BWV 773	634	11724
BWV 774	494	9882
BWV 775	443	9304
BWV 776	733	15978
BWV 777	547	17209
BWV 778	473	11103
BWV 779	598	11731
BWV 780	558	11995
BWV 781	439	9038
BWV 782	568	11306
BWV 783	685	15969
BWV 784	564	12250
BWV 785	592	16782
BWV 786	477	10407

Table 1: Number of datapoints (notes) and the number of discovered TECs in J. S. Bach's Two-Part Inventions.

$\langle P, T(P, D) \rangle$, where P is a maximally translatable pattern, and $T(P, D)$ is the set of translators for P in the dataset D. The worst-case running time of SIATEC is $O(kn^3)$.

Even for small datasets, the raw output of SIATEC can quickly become unmanageably large, as can be seen in Table 1. Furthermore, the patterns are diverse in size and structure, and on the whole are not readily intuitive. It would be straightforward to rank the discovered patterns based on a set of criteria; for example, to sort by pattern size $|P|$, or the number of pattern repetitions $|T(P, D)|$. However, such a simplistic approach presents two particular difficulties. Firstly, the method does not lend itself to any principled means of deciding how many of the most highly ranked patterns should be selected as being representative of the repetition in the dataset. Secondly, this method would preclude the ability to make inter-pattern judgments, that is, for the value of one pattern to influence the value of another, due to combinatorial explosion.

COSIATEC is one method for automatically identifying a subset of 'interesting' patterns from amongst the many patterns discovered by SIATEC [16, 14]. COSIATEC is designed to generate compressed representations of datasets by representing them in terms of highly repetitious subsets. The algorithm first runs SIATEC, generating a list of $\langle P, T(P, D) \rangle$ pairs, and then selects the best pattern based on heuristics. The algorithm then removes from the original dataset all the datapoints that are members of the occurrences of the chosen pattern P. The process continues until all the datapoints have been removed from the dataset. The resulting set of patterns are collectively termed a *cover* [19]. In this case, each datapoint is represented in a cover exactly once.

For each iteration of COSIATEC, the remaining patterns are evaluated according to the three heuristics of *coverage*, *compression ratio*, and *compactness*, and the most highly valued pattern is selected to become part of the resulting cover. These heuristics are also employed in the present work, and are defined below.

Although motivated by compression, COSIATEC has been shown to identify principal musical themes in pieces of music [16, 14]. This is explicable given the very nature of a musical theme, which will typically appear numerous times during a work, making it an ideal pattern for use in the encoding of a compressed representation. Therefore, COSIATEC offers a tidy solution to the difficulties of sifting through the output of SIATEC. The problem becomes one of generating optimal (finite) covers given particular heuristics. Furthermore, being a greedy algorithm, the generation of covers entails a degree of pattern co-dependency, since previously selected patterns will affect the outcome of later iterations.

3   An alternative selection algorithm

The approach to pattern discovery in the present work follows a similar strategy to that of COSIATEC. We again formulate the problem as one of cover generation, but explore possibilities created by shifting the emphasis away from purely optimising compression. The foremost difference in this approach is that we only apply SIATEC once—to initially process the entire dataset. Furthermore, we evaluate only once the musicological value of each pattern discovered by SIATEC. Therefore, the musicological value of each pattern is determined within the same initial context, prior to selection. The rational being that, compared with COSIATEC, the selection process in this case more closely relates to the process of musical listening, since listeners perceive patterns in a musical work in the context of all the notes. The resulting values, or weightings, are adjusted during subsequent iterations of the selection process, but only by the single varying factor of the number of currently uncovered datapoints that are covered by a particular pattern. This process is described below as an example of a weighted set-cover problem.

Also explored is the effect of loosening the constraint requiring that each datapoint only be covered once. This enables us to consider datapoints as belonging to multiple patterns within a single cover, creating the potential to infer structural relationships between repeated elements. A similar strategy could be pursued within COSIATEC, but only within the context of each iteration as covered datapoints are removed at each stage.

3.1   Generating set covers

The approach taken in this paper for generating covers from musical patterns discovered by SIATEC can be described in terms of the widely known NP-hard set-covering problem. Cormen et al. [5, pp. 1033–4] describe this problem:

> An instance $(X, \mathcal{F})$ of the ***set-covering problem*** consists of a finite set X and a family $\mathcal{F}$ of subsets of X, such that every element of X belongs to at least one subset in $\mathcal{F}$:

$$X = \bigcup_{S \in \mathcal{F}} S. \tag{1}$$

> We say that a subset $S \in \mathcal{F}$ ***covers*** its elements. The problem is to find a minimum-size subset $\mathcal{C} \subseteq \mathcal{F}$ whose members cover all of X.

$$X = \bigcup_{S \in \mathcal{C}} S. \tag{2}$$

In other words, the desired outcome of this optimisation problem is to find the smallest number of subsets in $\mathcal{F}$ that account for (cover) each element in X at least once.

The standard approach to set-cover utilises a greedy algorithm based on the heuristic of selecting at each stage the set S that covers the largest number of uncovered elements in X. Ties are broken randomly. This algorithm is the best known performing algorithm, achieving an approximation ratio of $(1 - o(1)) \ln n$ [7, p. 634].

In the context of SIATEC, X is equivalent to a dataset D. Similarly to the selection process of COSIATEC, we consider all instances of a pattern P in a dataset D as constituting a single subset of D. We use P_{TEC} in order to explicitly refer to a subset of D that is the union of each occurrence of P in D. Therefore, each subset $S \in \mathcal{F}$ is a P_{TEC}, and $\mathcal{F}$ is equivalent to the entire set of patterns (considering each as a P_{TEC}) discovered by SIATEC: $\mathcal{F} = \langle P_{\text{TEC}0}, P_{\text{TEC}1}, \ldots, P_{\text{TEC}n-1} \rangle$.

For our purpose, simply finding a minimum-sized subset $\mathcal{C} \subseteq \mathcal{F}$ does not adequately characterise the problem, since we require a means of specifying which patterns should be considered 'better' or 'worse' by the selection algorithm. Therefore, a more appropriate model for the problem is the equally well-known generalisation of the minimal set-cover problem: the minimum weighted set-cover problem [3]. Typically, a greedy algorithm is also adopted, except that covers are selected in the order that minimises the ratio of cover weight to number of elements covered.

To place the SIATEC cover problem in this context, it is necessary to attach weighting values to each of the discovered patterns in $\mathcal{F}$. This step is similar to the use of heuristics in COSIATEC, except that in this case the values are calculated only once, prior to the actual selection process. The higher a pattern scores according to a heuristic, the more relevant it is considered to be to the analysis. The heuristics used to calculate these values are discussed in the following section.

Contrary to the more typical formation of weighted set-cover problems, the selection process in this case seeks to maximise: $weight(P_{\mathrm{TEC}}) \times cover\text{-}ratio(P_{\mathrm{TEC}})$ where $weight$ is the musicological value of the pattern P_{TEC} predetermined by a set of evaluation heuristics, and $cover\text{-}ratio$ is the ratio between the number of uncovered datapoints in the dataset D that are members of the pattern P_{TEC}, to the total number of uncovered datapoints in D. The definition of cover-ratio here is based on the concept of *coverage* used in COSIATEC, which is defined as: 'the number of datapoints in the dataset that are members of occurrences of the pattern' [16, p. 7]. A minimum cover-ratio threshold, which must be exceeded for a pattern to become a member of the cover, has proved a useful parameter in the generation of set-covers. In practice, a minimum cover-ratio threshold of between five percent and twenty percent is the typical useful range. Higher values in this range are particularly useful in order to generate covers consisting of only a small number of patterns. High cover-ratio values may lead to not every datapoint in D being represented in the set-cover. However, this is not necessarily unsatisfactory, since not every single note in a piece of music may prove sufficiently salient to be accounted for by a typical listener, or even professional analysts.

Once it has been determined that a pattern should become a member of the set-cover, a final step is taken to determine whether a pattern should be considered a *primary* or *secondary* pattern. This step is simply intended to make the generated results more comprehensible for the human analyst, by attempting to group together 'similar' patterns. If a pattern is the first pattern to be selected, it is simply defined as primary. Each subsequently selected pattern is compared to each existing primary pattern in terms of the number of datapoints they commonly cover. This is in order to identify the primary pattern that is 'most similar' to the newly selected pattern, quantified in terms of overlapping coverage. If the proportion of commonly covered datapoints is greater than an arbitrarily defined threshold—50 percent in this case—then the newly selected pattern is declared a secondary pattern, and grouped together with the most similar primary pattern. If the newly selected pattern is not considered similar to any of the other primary patterns it is declared a primary pattern. Whether a pattern is defined as primary or secondary has no bearing on the actual selection process, it is purely a means of organising the selected patterns, as well as offering an estimation of the number of distinct musical ideas present in the work.

4   Pattern evaluation heuristics

As noted above, there may be many different forms of repetition in a piece of music. It is therefore necessary to establish an evaluation criterion in order to automate the extraction of the kinds of repetitions that are considered relevant to an analytical objective. The cover-ratio heuristic, introduced in the previous section, is one measure of a pattern's value, and is recomputed for every pattern for each iteration of the cover generation algorithm. Here we describe two further heuristics that are used to provide static, or absolute, measures of the value of a pattern.

Compression ratio is defined as 'the compression ratio that can be achieved by representing the set of points covered by all occurrences of a pattern by specifying just one occurrence of the pattern together with all the non-zero vectors by which the pattern is translatable within the dataset' [14, p. 13]. Compression ratio can, therefore, be calculated in terms of coverage:

$$\text{compression ratio} = \frac{\text{coverage}}{|P| + |T(P, D)| - 1} \tag{3}$$

Compression ratio is particularly useful for identifying large, non-overlapping patterns that have many occurrences in a dataset.

The second heuristic used is *compactness*, which is defined as 'the ratio of the number of points in the pattern to the number of points in the region spanned by the pattern' [14, p. 13]. This heuristic applies to each *occurrence* of a pattern P belonging to a TEC. Therefore, unlike compression-ratio which generates a single value for a TEC, compactness produces $|T(P, D)|$ values for each TEC. In order to arrive at a single value for a TEC, since the selection algorithm generates covers by selecting P_{TEC} subsets of D, the obvious approach is to use either the mean or maximum compactness value as the TEC weighting value. From a musical perspective, selecting the maximum pattern compactness value to determine the weighting of a

TEC can be justified on the principle that a significant musical theme will typically have at least one relatively prominent (compact) occurrence in a work.

As discussed in [16] and [14], the definition of 'region', for example, as a segment, bounding box, or convex hull, impacts on the computed compactness value. For our purposes, we employ the notion of a region as a segment, but also take into account the musical voicing within a pattern. Therefore, all the notes that occur inclusively between the first note onset of a pattern and the final offset are considered to belong to the region (segment) spanned by the pattern. However, unlike previous work, we calculate the ratio using only those notes in the region that are also members of the voices present in the pattern. This decision is based on the assumption that notes belonging to the musical voices present in a pattern are more likely to influence its perceptual salience, compared with notes belonging to other musical voices. This assumption is consistent with empirical findings related to melodic streaming [1, pp. 61–64]. The present definition of compactness, relying to a certain degree on specific musical concepts, is less generic than the original geometrical definition. However, it has proved to be the most satisfactorily performing variant in our study. Furthermore, given that our testing dataset consists of the fifteen J. S. Bach Inventions, the strict two-part texture of the music gives additional credence to the utilisation of voicing information in the selection process.

Patterns are initially assigned a default weight of one, and each heuristic is implemented such that it returns a normalised weighting value of between zero and one. Weighting combination is multiplicative. Therefore, if a pattern is rated most highly by each heuristic, it would retain the final value of one. If a pattern scores zero for any of the heuristics the final rating would also be zero.

The values generated by heuristics can also be scaled in a variety of ways. Most relevant to the current analysis is that values can be adjusted using either a linear or exponential mapping, and can also be given explicit minimum and maximum thresholds beyond which patterns are assigned a weighting of zero. Setting a threshold for a heuristic is particularly useful as a means of excluding a subset of patterns prior to the cover generation process. The removal of redundant subsets $S \in \mathcal{F}$ is common in the literature [2, p. 2]. The density of coverage present in $\mathcal{F}$ plays a significant role in determining the bounds of approximation [4], so from a practical standpoint, the prior reduction in the size of a set cover instance may lead to improved solutions.

5   Results

We have applied the cover generation algorithm and heuristics discussed above to J. S. Bach's Two-Part Inventions (BWV 772–786). Each piece was analysed using a range of different parameters for each heuristic. Individual pieces from this set: BWV 722, 724 and 725, are also the subject of analysis in [16] and [14]. Therefore, we focus attention on these three pieces in order to draw some comparison.

The covers generated by our system robustly included the same patterns as those discovered by COSIATEC. A minor deviation from this is in BWV 775. COSIATEC selects the opening two bars as constituting the most important pattern in work. Our system makes an almost identical selection, except that it does not include the very first note. Both interpretations are valid when considering the full score. There are indeed many occurrences of the pattern discovered by COSIATEC, however, there are also many additional occurrences where the first note is different. The difference between the two results is presumably a reflection of a different balance of emphasis given to the heuristics used to evaluate the patterns.

Figures 1 and 2 show a complete cover generated from BWV 772 using the following heuristic parameters:

- Cover-ratio (min: 0.2)

- Compression-ratio (min: 0.25, max 1.0)

- Compactness (min: 0.25, max 1.0)

There are 9035 TECs discovered by SIATEC in BWV 772. After calculating the weights of each TEC using compression-ratio and compactness, only 129 TECs have non-zero weights. This considerable reduction indicates that a very large proportion of the patterns discovered by SIATEC are not relevant to this particular analytical focus. Generating a cover with the relatively high cover-ratio threshold of 20 percent produces a cover consisting of only six patterns—three primary and three secondary. Setting a lower threshold in this case tends to increase the number of secondary patterns discovered.

Patterns 1 and 2, the first and second patterns discovered, are the inversion of the subject, and subject itself respectively. These patterns are the same as those discovered by COSIATEC, and are labeled as the subject of the work as analysed by [6, p. 10]. The two secondary patterns, 1.1 and 2.1, are both clearly subsets of their parent patterns. From an analytical perspective, the most interesting aspect of these patterns is how their individual pattern of occurrence *differs* from that of their parent, which may be described as an instance of *entanglement* [19]. This is particularly apparent for pattern 2.1 from bar 16 until the end, where many instances of the pattern overlap. As well as highlighting the high density of this simple descending three-note quaver pattern at the end of the piece, the change in the translations of pattern 2.1 in relation to pattern 2 suggest some sort of developmental change to the primary pattern. In fact, this change corresponds to the note that immediately follows an occurrence of pattern 2, which in these closing bars forms an interval of a 2nd. All previous occurrences, except for the occurrence preceding bar 9, are followed by a larger interval, most commonly a 5th.

It cannot really be argued that pattern 1.2 is a perceptually salient pattern when considered as a single occurrence. However, when taking into account the larger pattern that is formed by the overlapping occurrences, an important pattern emerges (Figure 3). Pattern 1.2 is indicative of the gap-fill pattern that accompanies pattern 1 in bars 3–5 and 11–13 (where it appears in the treble voice), and is also embedded in the structure of pattern 1 itself.

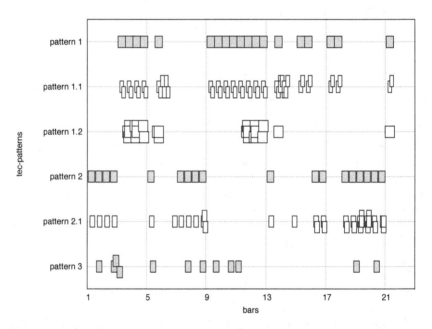

Figure 1: A schematic representation of the primary and secondary patterns selected from the SIATEC analysis of BWV 772. The filled boxes are primary patterns, the empty boxes are secondary patterns. Each box represents a pattern occurrence. To aid clarity, patterns that overlap are drawn alternately above and below the line.

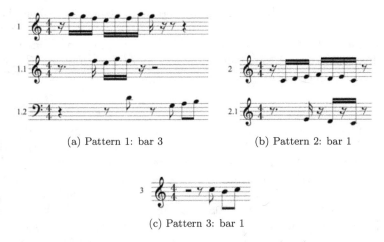

(a) Pattern 1: bar 3   (b) Pattern 2: bar 1

(c) Pattern 3: bar 1

Figure 2: The primary and secondary patterns selected from the SIATEC analysis of BWV 772.

Figure 3: Bars 3–4 from BWV 772. Square boxes above notes signify the beginning of an occurrence of pattern 1.2.

6   Future work

Applied examples from the literature present several variations on the greedy algorithm that have proved useful in particular domains, which may similarly be beneficial in our case. Marchiori and Steenbeek [13] describe the Enhanced Greedy algorithm, which has a more sophisticated heuristic for breaking ties when adding new covering sets of equal size to the solution. At each iteration the algorithm also checks for (and possibly removes) sets that become 'nearly' redundant in the solution due to the addition of new sets. The Iterated Enhanced Greedy algorithm is also described, in which a subset of the currently best (smallest) cover is used as an initial partial solution for a further iteration of the algorithm. Another approach that would also warrant empirical investigation in this context is the multiple weighted set cover problem, which is a further generalisation of the basic set-cover problem where events must be covered a specified minimum number of times [20]. Alternative approaches to the basic greedy algorithm, including approximate linear programming and exact branch and bound method, are discussed in [2].

7   Conclusion

An approach to automated musical analysis has been presented. The method is essentially a selection algorithm based on a set of heuristics that attempt to determine the quality of discovered patterns in terms of musical salience. The SIATEC algorithm is integral to the process, since it provides the initial set of discovered patterns from which the selection is made. Our method also owes much to the COSIATEC algorithm, which is also a means of selecting important patterns from the patterns discovered by SIATEC. The primary difference between our approach and COSIATEC is that our method is not based solely on the principle of optimising compression, but instead allows musicological principles to influence the outcome alongside information theoretic measures. As a result, we are able to select patterns that are deemed to be of musicological interest, but which may not lead to the generation of a complete or optimally compressed representation of the dataset, as is generated by COSIATEC. For example, we are able to select multiple patterns that share notes in common, but which have different patterns of occurrence within a piece. The ability to analyse the occurrences of closely related patterns within a work can provide interesting insight into the compositional treatment of thematic ideas.

There is still a great deal of work to be done in order to improve the quality and reliability of automated music analysis. However, even as it stands, our system opens up some very interesting possibilities for future work. Automated systems cannot currently hope to match the quality of analysis performed by professional musicologists, but do have an advantage of being able to process very large amounts of data. The ability to reliably isolate significant musical patterns and infer basic structural relationships between patterns from across a database of many thousands of pieces of music would create a rich source of musical knowledge, with exciting potential for a range of future research.

BIBLIOGRAPHY

[1] A. S. Bregman. *Auditory Scene Analysis: The Perceptual Organization of Sound.* The MIT Press, Cambridge, MA, USA, 1990.

[2] A. Caprara, M. Fischetti, and P. Toth. Algorithms for the set covering problem. Technical report, Technical Report No. OR-98-3. DEIS-Operations Research Group, 1998, 1998.

[3] V. Chvatal. A greedy heuristic for the set-covering problem. *Mathematics of Operations Research*, 4(3):233–235, 1979.

[4] K. L. Clarkson and K. Varadarajan. Improved approximation algorithms for geometric set cover. In *SCG '05: Proceedings of the twenty-first annual symposium on Computational geometry*, pages 135–141, New York, NY, USA, 2005. ACM.

[5] T. H. Cormen, C. E. Leiserson, R. L. Rivest, and C. Stein. *Introduction to Algorithms*. The MIT Press, Cambridge, MA, USA, 2nd edition, 2001.

[6] L. Dreyfus. *Bach and the Patterns of Invention*. Harvard University Press, Cambridge, MA, USA, 1996.

[7] U. Feige. A threshold of ln n for approximating set cover. *Journal of the ACM*, 45(4):634–652, July 1998.

[8] D. Huron. *Sweet Anticipation: Music and the psychology of expectation*. A Bradford Book MIT Press, Cambridge, MA, USA, 2006.

[9] R. M. Karp. Reducibility among combinatorial problems. In R. Miller and J. Thatcher, editors, *Complexity of computer computations*, pages 85–103. Plenum Press, New York, NY, USA, 1972.

[10] C. Krumhansl. Effects of perceptual organization and musical form on melodic expectancies. In M. Leman, editor, *Music, Gestalt, and Computing: Studies in Cognitive and Systematic Musicology*, volume 1317 of *LNAI*, pages 294–320. Springer-Verlag, Heidelberg, Germany, 1997.

[11] K. Lemström and A. Pienimäki. On comparing edit distance and geometric frameworks in content-based retrieval of symbolically encoded polyphonic music. *Musicæ Scientiæ*, Discussion Forum 4A:135–152, 2007.

[12] F. Lerdahl and R. Jackendoff. *A Generative Theory of Tonal Music*. MIT Press, Cambridge, MA, USA, 1999 edition, 1983.

[13] E. Marchiori and A. Steenbeek. An iterated heuristic algorithm for the set covering problem. In K. Mehlhorn, editor, *Proceedings of the Workshop on Algorithm Engineering*, pages 155–166, 1998.

[14] D. Meredith. Point-set algorithms for pattern discovery and pattern matching in music. In T. Crawford and R. C. Veltkamp, editors, *Proceedings of the Dagstuhl Seminar on Content-Based Retrieval*, number 06171 in Dagstuhl Seminar Proceedings, Dagstuhl, Germany, April 2006. Internationales Begegnungs- und Forschungszentrum fuer Informatik (IBFI), Schloss Dagstuhl.

[15] D. Meredith, K. Lemström, and G. A. Wiggins. Algorithms for discovering repeated patterns in multidimensional representations of polyphonic music. *Journal of New Music Research*, 31(4):321–345, 2002.

[16] D. Meredith, K. Lemström, and G. A. Wiggins. Algorithms for discovering repeated patterns in multidimensional representations of polyphonic music. In *Cambridge Music Processing Colloquium 2003*, Cambridge, UK, 2003. Department of Engineering, University of Cambridge.

[17] J.-J. Nattiez. *Musicologie générale et sémiologue (Music and Discourse: Towards a Semiology of Music (1990), trans. by Carolyn Abbate)*. Princeton University Press, Princeton NJ, USA, 1987.

[18] E. Ukkonen, K. Lemström, and V. Mäkinen. Sweepline the music! In R. Klein, H.-W. Six, and L. Wegner, editors, *Computer Science in Perspective*, volume 2598 of *LNCS*, pages 330–342. Springer-Verlag, Berlin, Germany, 2003.

[19] G. A. Wiggins, K. Lemström, and D. Meredith. Sia(m)ese: An algorithm for transposition invariant, polyphonic content-based music retrieval. In M. Fingerhut, editor, *Proceedings of the 3rd International Conference on Music Information Retrieval*, pages 283–284, 2002.

[20] J. Yang and J. Y.-T. Leung. A generalization of the weighted set covering problem. *Naval Research Logistics*, 52(2):142–149, 2005.

Jamie Forth
Goldsmiths, University of London,
London, SE14 6NW, UK
Email: j.forth@gold.ac.uk

Geraint A. Wiggins
Goldsmiths, University of London,
London, SE14 6NW, UK
Email: g.wiggins@gold.ac.uk

An Improved Algorithm for Rhythm Recognition in Musical Sequences[1]

MARCIN KUBICA AND TOMASZ WALEŃ

ABSTRACT. One of the fundamental problems in musical analysis is the problem of classifying songs according to their rhythm. A song is represented by a sequence of numbers denoting duration of notes, and a rhythm is represented by a sequence of symbols Q ('Quick') and S ('Slow'), corresponding to relative durations of 'steps', such that $S = 2Q$.

In this paper we study the problem of covering a given musical sequence with a given rhythm. We present an efficient algorithm for finding the maximum-length substring coverable by the given rhythm, running in $O(n \cdot (\log H + \log m))$ time, where n is the length of the musical sequence, m is the length of the rhythm and H is the maximum duration of a note.

1 Introduction

Automated musical analysis has been studied extensively in the literature [8, 9, 11]. Music can be classified according to many characteristics, e.g. melody, rhythm, genre. These characteristics are natural for human beings, but their automated recognition is a hard task, especially that there is no full agreement on the formal definition of these characteristics.

Many problems arising there correspond to fundamental problems in combinatorial algorithmics and text processing, e.g. [5, 6]. One of the types of such problems are those related to rhythm analysis [1, 2, 3, 4].

In this paper we focus on classification of musical sequences according to their rhythm. We follow the framework presented in [2, 3, 4]. More precisely, we study the problem of finding a maximum-length substring of a musical sequence, that can be covered with occurrences of a given rhythm pattern. This problem was introduced in [3, 4] and a solution presented there has $O(nm \frac{\log H}{w})$ time complexity (where n is the length of the analysed musical sequence, m is the length of the matched rhythm, H is the longest interval

[1]This research was supported by the Polish Ministry of Science and Higher Education under grant N206 004 32/0806.

between consecutive rhythmic events, and w is the size of the machine word). The solution presented in this paper improves it, having $O(n \cdot (\log H + \log m))$ time complexity.

The structure of this paper is as follows: In Section 2 basic definitions are given, together with a formal definition of the problem of finding the maximum-length substring coverable by a given rhythm. In Section 3 we present an algorithm solving this problem in $O(n \cdot (\log H + \log m))$ time. Finally, we conclude in Section 4.

2 Preliminaries

In this section we present basic definitions. Generally, we follow the framework presented in [2, 3, 4]. A musical sequence is a sequence of notes (or their concords) characterised by their respective pitch and duration. For the purpose of rhythm recognition, we can focus on the intervals between the onset of the notes only and abstract from the pitch of tones or their concords. Hence, we will represent musical sequences as sequences of positive integers denoting relative intervals between rhythmical events, consecutive notes or their concords. We also assume that the range of these integers is such that each of them can be stored in a single machine word. Taking into account durations of intervals between notes in actual pieces of music, such an assumption is quite natural and the presented model is more general than required for musical applications. Formally, a *musical sequence* (or a *musical string*) t is a string $t = t[1] \ldots t[n]$, where $t[i] \in \mathbb{N}^+$, for $i = 1, 2, \ldots, n$. We also define H as $H = \max\{t[i] : 1 \le i \le n\}$.

For example, $(2, 7, 1, 10)$ and $(10, 35, 5, 50)$ are valid musical sequences. Moreover, the latter one can be obtained from the first one by multiplying the elements by 5. So, the only difference between them is the actual unit duration. Since the integers represent relative intervals between notes, such musical sequences are indistinguishable in rhythm analysis.

A *rhythm* is a sequence of steps; fast (denoted by Q) or slow (denoted by S). Steps Q and S represent durations, such that the duration of S is double the duration of Q. The actual durations of Q and S are not fixed, which makes the problem of rhythm recognition more complex than classical text pattern matching. Formally, a rhythm r is a string $r = r[1] \ldots r[m]$, where $r[i] \in \{Q, S\}$, containing at least one symbol S.

Let q be a positive integer, $q \in \mathbb{N}^+$, representing the duration of Q. Q matches a single interval with duration q, or a musical subsequence with total duration q. Formally, we say that Q q-*matches* a substring of a musical sequence $t[i]t[i+1] \ldots t[j]$ (for $1 \le i \le j \le n$), when $t[i] + t[i+1] + \cdots + t[j] = q$. Moreover, if $i = j$, that is Q q-matches $t[i]$, we say that a q-match is *solid*.

S matches a single interval with duration $2q$, or a sequence of notes that can be split into two sequences q-matching Q. Formally, we say that S q-*matches* a substring of a musical sequence $t[i]t[i+1]\ldots t[j]$ (for $1 \le i \le j \le n$), when:

- $i = j$ and $t[i] = 2q$ (in such a case we also say, that a q-match is *solid*),
  or

- for some k, $i < k \le j$, we have $t[i]+\cdots+t[k-1] = t[k]+\cdots+t[j] = q$.

If the value of q is known from the context, we will simply write 'match' instead of 'q-match'.

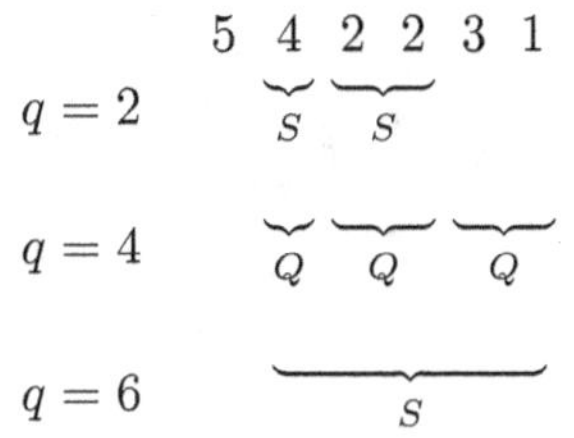

$$5 \quad 4 \quad 2 \quad 2 \quad 3 \quad 1 \quad 2$$

$$q = 2 \qquad \underbrace{\quad}_{S} \ \underbrace{\quad}_{S}$$

$$q = 4 \qquad \underbrace{\quad}_{Q} \ \underbrace{\quad}_{Q} \ \underbrace{\quad}_{Q}$$

$$q = 6 \qquad \underbrace{\qquad\qquad}_{S}$$

Figure 1. q-matching of Q and S.

Let us have a look at the example in Fig. 1. The musical sequence under consideration is $(5, 4, 2, 2, 3, 1, 2)$. For $q = 2$ we have two 2-matches of S: (4) (which is a solid one) and (2, 2). Please note, that there is no 2-match of S for (3, 1) — although the total duration is 4, it cannot be split into two parts, each of duration 2. For $q = 4$ we have three 4-matches of Q: the exact same substrings that 2-match S and additionally the substring (3, 1). Again, 4 is a solid match. For $q = 6$ we have only one 6-match of S: (4, 2, 2, 3, 1).

A rhythm r matches a musical subsequence if it can be split into parts matching consecutive elements of r. Formally, we say that $r = r[1]r[2]\ldots r[m]$ q-matches a substring of a musical sequence $t[i]t[i+1]\ldots t[j]$ (for $1 \le i \le j \le n$), if:

- there exists a sequence of indices $i = k_1 < k_2 < \cdots < k_{m+1} = j+1$, such that $r[l]$ q-matches $t[k_l]\ldots t[k_{l+1}-1]$, for $l = 1, 2, \ldots, m$,

- at least one of these matches is a solid match of S.

A match of a rhythm is also called its *occurrence*. If in a q-match of the rhythm all occurrences of S are solid, we call such a q-match (together with its position) *basic*.

Please consider a q-occurrence of some rhythm r. Let r' be a rhythm obtained from r by replacing all steps S with non-solid occurrences with QQ. Please note, that r' q-occurs in the exact same place and its q-occurrence is basic. We can formalize it, by introducing a notion of 'matching' between rhythms. We say that a rhythm $r \in \{Q, S\}^*$ *matches* sequence $r'' \in \{Q, S\}^*$ iff r'' can be obtained from r by replacing some (but not all) S-s by QQ. For example, the rhythm $r = QSS$ matches sequence $\pi_1 = QSQQ$, but does not match $\pi_2 = QQSQ$.

If for a given q the occurrences of the rhythm cover the whole musical sequence, we say that the rhythm q-*covers* the musical sequence. If there exists q for which the rhythm q-covers the musical sequence, we simply say that it *covers* the musical sequence.

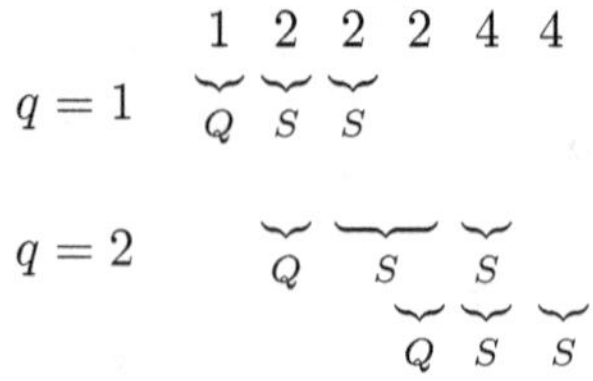

Figure 2. q-matching of rhythm QSS.

Let us have a look at the example in Fig. 2. The musical sequence under consideration is $(1, 2, 2, 2, 4, 4)$. Rhythm QSS has three occurrences. One for $q = 1$, namely $(1, 2, 2)$, and two for $q = 2$, namely $(2, 2, 2, 4)$ and $(2, 4, 4)$. Please note, that although the three occurrences of QSS cover the whole sequence, it is not q-covered by QSS, since the occurrences are for two different values of q. But, QSS 2-covers $(2, 2, 2, 4, 4)$.

In this paper we consider the problem of finding the longest substring of the given musical sequence, covered by the given rhythm. It is formally defined as follows:

PROBLEM 1 (Maximum Coverability Problem). We are given a musical sequence $t = t[1] \ldots t[n]$ ($t[i] \in \mathbb{N}^+$) and a rhythm $r = r[1] \ldots r[m]$ ($r[i] \in \{Q, S\}$). The maximum coverability problem (MCP) is to find the longest substring $t[i..j]$ of t that is covered by r (that is, there exists q, for which r q-covers the substring).

We can safely assume that $m \leq n$, since otherwise there cannot be any occurrence of r in t. We can also assume that $\log H = O(\log n)$, since otherwise the interval between notes would not fit into a single machine word, and the cost of arithmetical operations could not be considered constant.

3   Maximal coverability algorithm

In this section, we present an efficient algorithm for the MCP problem. Let $t = t[1] \ldots t[n]$ be a given musical sequence and $r = r[1] \ldots r[m]$ be a given rhythm. Generally, we follow the scheme defined in [2, 3, 4]:

1. First, for all possible values of q, we find maximum-length sequences with basic q-occurrences in t.

2. Then, we identify occurrences of pattern r, based on the results of the previous step.

3. Finally, for all possible values of q, we identify regions covered by occurrences of pattern r, and find the maximum-length substring that is q-covered.

The first step of the algorithm is the same as in [2, 3, 4]. The main difference between the algorithm presented here and in [2, 3, 4] is in the second step. Fischer and Paterson [7] observed that many string matching problems can be solved using FFT to compute Boolean convolutions. We also use Boolean convolutions to efficiently match r against solid occurrences of S surrounded by sequences of Q. We also improve the analysis of the first step, which leads to a proof of better total time complexity of the algorithm. In the following subsections we describe the consecutive steps of the algorithm.

3.1   Transform

The first step of the algorithm is the same as in [4]. We give a detailed description for clarity. The rhythm r must contain at least one S, and each occurrence of r in t must contain at least one solid occurrence of S. Each non-solid occurrence of S is equivalent to an occurrence of QQ. For fixed q and such position i that $t[i] = 2q$, we can construct maximum-length sequence π that has a basic q-occurrence in t containing position i. Moreover, π contains S corresponding to $t[i]$. The sequence π contains enough information to find (within its q-occurrence) any q-occurrence of any rhythm. Simply, at least one solid occurrence of S must coincide with S in π, non-solid occurrences of S coincide with QQ in π and all Q-s coincide with Q-s in π.

Therefore, for a given duration q, we first find all maximum-length sequences with basic q-occurrences in t. Such sequences must have the form $Q^{i_1} S Q^{i_2} S \ldots S Q^{i_k}$, where $k > 1$, and $i_l \geq 0$ (for $1 \leq l \leq k$).

Let us consider two such consecutive maximum-length q-occurrences of sequences. They can either be disjoint (see Fig. 3.a) or can overlap (see Fig. 3.b). However, in the latter case only sequences of Q-s can overlap, and their positions cannot coincide.

Please note, that in the first case both q-occurrences cannot be in the same substring q-covered by any rhythm. In the latter case both q-occurrences can be in the same substring q-covered by rhythm r, but when looking for q-occurrences of r in t we can consider them separately.

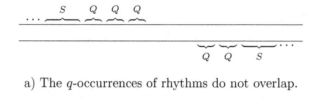

a) The q-occurrences of rhythms do not overlap.

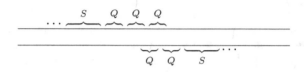

b) The q-occurrences of rhythms overlap,
but positions of Q-s do not coincide.

Figure 3. Possible relative layouts of maximum-length basic occurrences of rhythms.

Figure 4 presents an algorithm *Transform* finding, for a given q, all maximum-length sequences with basic q-occurrences in t. The sequences are constructed in a greedy manner. First, we find the solid occurrence of S (using $O(n \log H)$ time preprocessing, we can construct collections of occurrences of S, for all possible values of q). Then, we extend it to the left with a maximum-length sequence of Q-s. Then, we extend it to the right, looking for q-occurrences of Q and/or solid q-occurrences of S. Since the basic occurrences of sequences can only overlap on Q-s, we continue construction of the next basic q-occurrence looking for the first solid q-occurrence of S after the previously constructed basic q-occurrence.

Transform returns the set of maximum-length sequences together with positions of their basic q-occurrences. However, for clarity of the presentation, we skip the positions and construct just the set of rhythms.

LEMMA 2. *The total running time of the algorithm* Transform *(for all values of q) is $O(n \log H)$, where $H = \max\{t[i] : 1 \le i \le n\}$.*

Proof. The lemma has been proved in [4]. The main argument is that each element $t[i]$ can be inspected by the algorithm at most $O(\log H)$ times. ∎

---

function Extend($t[1..n], i, q, dir$)

---

1   $sum = 0$
2   **while** $sum < q$ **and** $i \geq 1$ **and** $i \leq n$ **do**
3   **if** $sum + t[i] = q$ **then**   **return** i
4      $sum = sum + t[i]$
5   **if** $dir = left$ **then** $i = i - 1$ **else** $i = i + 1$
6   **end**
7   **if** $dir = left$ **then return** $-\infty$ **else return** ∞

---

function Transform($t[1..n], q$)

---

1   $\Pi_q = \emptyset$
2   $last = -\infty$
3   **for** $k \in \{i : t[i] = 2q\}$ $\{in\ ascending\ order\}$ **do**
4   **if** $k \leq last$ **then continue**
5      $\pi = \text{`}S\text{'}; \ last = k$
6      $j = k$; { extend pattern to the left }
7   **while** $j > 1$ **do**
8         $j = Extend(t, j - 1, q, left)$
9   **if** $j \geq 1$ **then** $\pi = \text{`}Q\text{'}{+}\pi$
10   **end**
11      $j = k$; { extend pattern to the right }
12   **while** $j < n$ **do**
13   **if** $t[j + 1] = 2q$ **then**
14            $\pi = \pi + \text{`}S\text{'}; \ j = j + 1 \ ; \ last = j$
15   **else**
16            $j = Extend(t, j + 1, q, right)$
17   **if** $j \leq n$ **then** $\pi = \pi + \text{`}Q\text{'}$
18   **end**
19   **end**
20   insert rhythm π to the set Π_q;
21   **end**
22   **return** Π_q

Figure 4. Algorithm Transform.

As we have mentioned in Section 2, we can assume that $\log H = O(\log n)$. Otherwise, the assumption that each value $t[i]$ fits in a single machine word would not be justified, and one should take into account the cost of arithmetic operations.

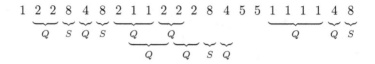

Figure 5. Example application of algorithm Transform, for $q = 4$, we obtain $\Pi_q = \{QSQSQQ,\ QQSQ,\ QQS\}$.

Figure 5 presents the results of application of algorithm *Transform* for $q = 4$ and $t = (1, 2, 2, 8, 4, 8, 2, 1, 1, 2, 2, 2, 8, 4, 5, 5, 1, 1, 1, 1, 4, 8)$. The resulting maximum-length sequences with basic q-matches are: $QSQSQQ$, $QQSQ$ and QQS. Please note, that the occurrences of the first two sequences overlap, but only on sequences of Q-s.

In [4] it has been proved, that the total length of rhythms computed by the algorithm *Transform* (for all values of q) is $O(n \log H)$. We will improve this, proving the following lemma:

LEMMA 3. *The total length of rhythms computed by the algorithm* Transform, *for all possible values of q, is not greater than $5n$.*

Proof. The total number of characters S in rhythms computed by the algorithm is exactly n, since each element $t[i]$ can be interpreted as a solid occurrence of S only for $q = t[i]/2$. We will show, that the number of characters Q can be bounded by $4n$.

Let us assign to each element $t[i]$ a weight w_i and lists L_i and R_i. Initially every element has weight 0 and the lists are empty. Whenever the algorithm produces Q matching elements $t[i, \ldots, j]$, we increase weight of these elements, namely for $i \leq k \leq j$, we increase the weight w_k by $t[k]/q$. Note, that the total increase of weight for a single character Q is 1. So, the total number of characters Q generated by the algorithm (for all values of q) is equal to the sum of weights $\sum_{i=1,\ldots,n} w_i$.

Let us consider a single invocation of *Transform* for a given value of q. Each element $t[i]$ can be matched against Q at most twice — once when some rhythm is extended to the right, and then when the following rhythm is extended to the left. So, its weight can increase by at most $2t[k]/q$. Whenever the algorithm *Transform* increases the weight w_k when extending some rhythm to the left (to the right) we add q to the list L_k (R_k).

Now we will prove that the final value of w_i is not greater than 4, for every $i = 1, \ldots, n$. Let us investigate element $t[i]$ with weight w_i and lists

L_i and R_i. For simplicity, we assume that the lists are sorted in ascending order. Let $R_i = q_1, \ldots, q_k$, where $q_j < q_{j+1}$. Since $t[i]$ is a part of a sequence matching Q, we have $t[i] \leq q_1$. Let us consider values q_j and q_{j+1} (for $1 \leq j < k$). These values correspond to some q-occurrences with a solid match of S to the left from position i. Please note, that the solid match of S corresponding to q_{j+1} must be further to the left than one corresponding to q_j. So, the solid match of S corresponding to q_j must be a part of a sequence (q_{j+1})-matching Q (see Fig. 6). Therefore $2q_j \leq q_{j+1}$, and $2^{j-1}q_1 \leq q_j$. Hence:

$$\sum_{q \in L_i} \frac{t[i]}{q} \leq \sum_{i=0,1,\ldots} \frac{t[i]}{q_1 \cdot 2^i} \leq \frac{2t[i]}{q_1} \leq 2$$

Using the same argumentation, we can prove that:

$$\sum_{q \in R_i} \frac{t[i]}{q} \leq 2$$

The total weight w_i obtained by element $t[i]$ (after processing all values of q) can be computed as:

$$w_i = \sum_{q \in L_i} \frac{t[i]}{q} + \sum_{q \in R_i} \frac{t[i]}{q} \leq 4$$

Concluding, $\sum_{i=1 \ldots, n} w_i \leq 4n$, and the total number of characters Q and S in all the rhythms generated by the algorithm *Transform* is not greater than $5n$. ∎

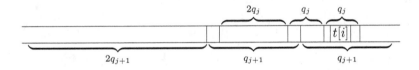

Figure 6. Relation between different values of q for which $t[i]$ is a part of a sequence matching Q.

3.2   Finding occurrences of the pattern

After the computation of sets Π_q, we are ready for the next stage of the algorithm — finding occurrences of the pattern r in strings $\pi \in \Pi_q$. Please note, that in this step, both r and π contain only the characters Q and S.

Let us define function *Encode*, which transforms strings from $\{Q, S\}^*$ into binary strings. The function *Encode* replaces each character Q by 000 and each character S by 110011. For example, $Encode(SQQ) = 110011000000$.

The main difference between traditional pattern matching and finding occurrences of rhythm r in sequence π is the possible length of the occurrences. If the rhythm r occurs in π on positions $a..b$, we can bound the length of the occurrence by $|r| \leq |b - a + 1| < 2|r|$. The main idea of the function *Encode* is to remove that difficulty and ensure that each occurrence has the same length. Let $\pi' = Encode(\pi)$, $r' = Encode(r)$ and let the rhythm r occur in π on positions $a..b$, which corresponds to positions $a'..b'$ in π'. We have $|b' - a' + 1| = |r'|$.

We say that j is a valid position if substring $\pi'[j..(j+|r'|-1)]$ corresponds to some substring $\pi[a..b]$. Then j is a valid position if and only if:

- $j \mod 3 = 1$ and $1 \leq j \leq |\pi'| - |r'| + 1$, and

- $\pi'[j..(j+2)] \neq 011$, and

- $\pi'[(j + |r'| - 3)..(j + |r'| - 1)] \neq 110$.

Instead of searching for occurrences of r in π, we will look for the occurrences of r' in π'. Once again we will have to define the notion of occurrence. We have an occurrence of r' in π' on the positions $j..(j + |r'| - 1)$ iff:

- j is a valid position,

- $\pi'[j + k - 1] = 1 \implies r'[k] = 1$, for $1 \leq k \leq |r'|$, and

- $\pi'[j + k - 1] = 1$ for some $1 \leq k \leq |r'|$.

Figure 7 shows the algorithm *FindOccurrences* encoding r and π as r' and π', and finding all occurrences of r' in π'. Let us denote by $u(j) = \sum_{i=1}^{|r'|} \pi'[j+i-1]$ and by $s(j) = \sum_{i=1}^{|r'|} \pi'[j+i-1]r'[i]$. Please note, that we have an occurrence of r' in π' on the positions $j..(j + |r'| - 1)$, if and only if:

- j is a valid position,

- $s(j) = u(j)$, and

- $s(j) > 0$.

Hence, it is enough to calculate values $s(j)$ and $u(j)$, and compare them for valid positions j.

LEMMA 4. *The total running time of the second phase of the presented algorithm is $O(n \log m)$.*

---

function FindOccurrences(π,r)

1 let $\pi' = Encode(\pi)$
2 let $r' = Encode(r)$
3 compute $u(j) = \sum_{i=1}^{|r'|} \pi'[j + i - 1]$
4 compute $s(j) = \sum_{i=1}^{|r'|} \pi'[j + i - 1]r'[i]$ using Boolean convolutions
5 output match for those j's where j is a valid position, $s(j) = u(j)$ and $s(j) > 0$

Figure 7. Algorithm FindOccurrences.

Proof. The running time of algorithm *FindOccurrences* is $O(|\pi| \log |r|)$, since the most time consuming step of this algorithm is the computation of values $s(j)$ using Boolean convolutions [12] and FFT [10]. From Lemma 3 we know that the total length of all patterns π generated for all values of q is $O(n)$. Hence, the total time required to find all occurrences of r in t is $O(n \log m)$. ∎

3.3   Identifying regions covered by occurrences of r

In the previous phase of the algorithm, we have identified all occurrences of rhythm r' in π'. Each such occurrence corresponds to an occurrence of r in t. We can easily construct the sets of occurrences:

$$Occ_q = \{(i, j) : \text{rhythm } r \text{ } q\text{-matches } t[i..j]\}$$

The last part of the algorithm is finding the longest interval covered by occurrences from Occ_q. This can be easily done by sorting intervals of Occ_q by their left endpoints and then scanning the obtained sequence.

LEMMA 5. *The total running time of the last phase of the presented algorithm is $O(n)$.*

Proof. From Lemma 3 and the definition of the *Encoding* function, we know that the total length of all sequences π' (generated for all values of q) is $O(n)$. Hence, the total number of intervals in all sets Occ_q is also $O(n)$. Since endpoints of intervals are from the range $1..|t|$, we can use radix sort for sorting Occ_q. Hence, the total running time of this phase of the algorithm is $O(n)$. ∎

THEOREM 6. *The total running time of the presented algorithm is $O(n \cdot (\log H + \log m))$.*

Proof. It is an immediate consequence of Lemmas 2, 4 and 5. ∎

4   Conclusions

In this paper we have presented an improved algorithm for identifying occurrences of given rhythm in musical sequences. Our algorithm requires $O(n \cdot (\log H + \log m))$ time, improving the previous algorithm with complexity $O(nm\frac{\log H}{w})$ (where w is the size of the machine word) proposed in [4]. The improvement is obtained by more detailed analysis and use of Boolean convolutions for identifying occurrences.

Although, from the theoretical point of view it is the improvement, in practice H, m and w can be considered constant. From such a point of view, both algorithms run in linear time.

Acknowledgements   The authors would like to thank anonymous reviewers for the time and effort spent on improving the contents of the paper.

BIBLIOGRAPHY

[1] E. Cambouropoulos, M. Crochemore, C. S. Iliopoulos, M. Mohamed, and M.-F. Sagot. A pattern extraction algorithm for abstract melodic representations that allow partial overlapping of intervallic categories. In *ISMIR*, pages 167–174, 2005.

[2] A. L. P. Chen, C. S. Iliopoulos, S. Michalakopoulos, and M. S. Rahman. Implementation of algorithms to classify musical texts according to rhythms. In C. Spyridis, A. Georgaki, G. Kouroupetroglou, and C. Anagnostopoulou, editors, *Proceedings of the 4th Sound and Music Computing Conference, SMC'07*, pages 134–141, 2007.

[3] M. Christodoulakis, C. S. Iliopoulos, M. S. Rahman, and W. F. Smyth. Song classifications for dancing. In J. Holub and J. Zdárek, editors, *Stringology*, pages 41–48. Department of Computer Science and Engineering, Faculty of Electrical Engineering, Czech Technical University, 2006.

[4] M. Christodoulakis, C. S. Iliopoulos, M. S. Rahman, and W. F. Smyth. Identifying rhythms in musical texts. *International Journal of Foundations of Computer Science*, 19(1):37–51, 2008.

[5] T. Crawford, C. S. Iliopoulos, and R. Raman. String matching techniques for musical similarity and melodic recognition. *Computing in Musicology*, 11:73–100, 1998.

[6] M. Crochemore, C. Iliopoulos, T. Lecroq, and Y. Pinzn. Approximate string matching in musical sequences. In *Proc. PSC 2001*, pages 26–36, 2001.

[7] M. Fischer and M. Paterson. String matching and other products. In *SIAM-AMS Proc.*, volume 7, pages 113–125, 1974.

[8] C. Iliopoulos, K. Lemström, M. Niyad, and Y. Pinzon. Evolution of musical motifs in polyphonic passages. In *Proc. AISB'2002 Symposium on AI and Creativity in Arts and Science*, pages 67–75, London, United Kingdom, 2002.

[9] K. Lemström and P. Laine. Musical information retrieval using musical parameters. In *Proc. 1998 International Computer Music Conference (ICMC'98)*, pages 341–348, Ann Arbor, USA, 1998.

[10] W. H. Press, B. P. Flannery, S. A. Teukolsky, and W. T. Vetterling. *Numerical Recipes: The Art of Scientific Computing*. Cambridge University Press, 1986.

[11] I. Shmulevich, O. Yli-Harja, E. Coyle, D.-J. Povel, and K. Lemström. Perceptual issues in music pattern recognition — complexity of rhythm and key finding. In *Proc. AISB'99 Symposium on Musical Creativity*, pages 64–69, Edinburgh, United Kingdom, 1999.

[12] M. Werman. Fast convolution. In *WSCG*, 2003.

Marcin Kubica
Institute of Informatics
University of Warsaw
Banacha 2, 02–097 Warszawa, Poland
Email: kubica@mimuw.edu.pl

Tomasz Waleń
Institute of Informatics
University of Warsaw
Banacha 2, 02–097 Warszawa, Poland
Email: walen@mimuw.edu.pl

Crochemore Sets

ALAN GIBBONS AND PAUL SANT

ABSTRACT. We introduce the notion of *Crochemore Sets* and provide counting theorems related to them. Their combinatorial richness is illustrated through connections made to *finite automata* and to *graph colourings*. The relationship we make with the famous *four colour problem of planar maps* adds to their innate interest.

1 Introduction

As far as we know, Maxime Crochemore has never worked on the eponymous subject of this paper. This dedication, to a fine man, stakes out new territory for him in even wilder regions of *stringology* than his extensive and outstanding work has yet encompassed.

A *Crochemore Set of degree* n, for some integer $n \geq 1$, has 2^{n-1} strings each of length n over a ternary alphabet $T = \{a, b, c\}$. We parameterise each Crochemore Set, $CS(G, D)$, by a *generator*, $G \in \{a, b, c\}$, and a sequence of digits, D, which generates the set from G. We first define Crochemore Sets inductively. The smallest sets are:

$$CS(a, \varnothing) = \{a\}, CS(b, \varnothing) = \{b\} \text{ and } CS(c, \varnothing) = \{c\}.$$

Here, in each case, D is the empty sequence and $n = 1$. For $n > 1$, D is a sequence of $(n-1)$ digits. We construct $CS(a, (d_1, d_2, \ldots, d_n))$ from $CS(a, (d_1, d_2, \ldots, d_{n-1}))$ as follows. Let S be any string in $CS(a, (d_1, d_2, \ldots, d_{n-1}))$ and let x be the (d_n)th character of S. Now let yz and zy be the two 2-character strings constructible from those distinct elements of T that are not x. From each S, we construct two strings for $CS(a, (d_1, d_2, \ldots, d_n))$, the first is obtained from S by replacing the (d_n)th character by yz and the second string is obtained from S by replacing the (d_n)th character by zy. Clearly, we require that $0 < d_n \leq n$. Some examples of small Crochemore Sets follow:

$$CS(a, (1)) = \{bc, cb\} \qquad CS(a, (1, 1)) = \{acc, cac, abb, bab\}$$
$$CS(a, (1, 2)) = \{bab, bba, cac, cca\} \qquad CS(b, (1, 2)) = \{aba, aab, cbc, ccb\}$$
$$CS(a, (1, 2, 1)) = CS(a, (1, 1, 3)) =$$
$$\{acab, acba, caab, caba, abac, abca, baac, baca\}$$

The third line of these examples shows that, as we have introduced it, our notation may specify the same set in more than one way. This can be avoided by insisting on *normalised* notation in which the digits of D form a *non-decreasing* sequence. In other words, $d_i \leq d_{i+1}$, for $0 < i < n$. It is easy to see that every CS can be specified in normalised notation. We use this notation for the rest of the paper. The examples also illustrate the following facts:

- Any two CS's with different generators are *disjoint*.

- Any two CS's of the same degree and same generator have a *non-empty* intersection.

We omit any proof of the first of these facts which is an easy by-product of Theorem 1. The second fact, is, however, a different matter. We provide a proof in the final section of this paper by showing its equivalence to the famous *Four Colour Problem of Planar Maps* (FCP). The implication is that a proof independent of FCP would be extremely difficult. In fact, it would also provide a new proof for FCP and this would be of universal interest.

In contrast to the inductive definition for a CS above, we now provide a recurrence relationship. Suppose that our *Crochmore Set* is $CS(a, D)$, where $D = (d_1, d_2, \ldots, d_k)$. If it exists, let d_i be the first digit of D for which $i \neq 1$ and $d_i = i$. Then from D, we define two other sequences, D_1 and D_2 as follows:

$$D_1 = \text{if } (i{=}2) \text{ then } \varnothing$$
$$\qquad \text{otherwise \{including } if \text{ (i does not exist)\} } (d_2, d_3, \ldots, d_{i-1})$$
$$D_2 = \text{if } (i \text{ does not exist}) \text{ then } \varnothing$$
$$\qquad \text{otherwise } ((d_i - i + 1), (d_{i+1} - i + 1), \ldots, (d_k - 1 + 1))$$

We define a Cartesian product of sets as follows. For sets of strings A and B, the set $A \times B$ contains precisely all possible strings obtained by concatenating a string of A (as a prefix) with a string of B (as a suffix). We can now state the following recurrence relationship:

$$CS(a, D) = (CS(b, D_1) \times CS(c, D_2)) \cup (CS(c, D_1) \times CS(b, D_2))$$

where the smallest *Crochmore Sets*, $CS(a, \varnothing)$, $CS(b, \varnothing)$ and $CS(c, \varnothing)$ were defined earlier. If the veracity of this recurrence is not immediately obvious, it will become so by the end of the paper.

In the following section we provide two basic counting theorems related to *Crochemore Sets*. Then, in the ensuing sections, we make connections to finite automata, regular languages and tree colourings, culminating in a proof that any two CS's of the same degree and same generator have a non-empty intersection.

2   Counting theorems for Crochemore Sets

Here we present two theorems. The first essentially counts the number of distinct strings over $\{a, b, c\}$ which belong to at least one *Crochemore Set*. We shall see that, asymptotically, a single generator (a, b or c) will account for one quarter of all possible distinct strings over the ternary alphabet T. The second theorem counts the number of distinct *Crochemore Sets* of degree n.

THEOREM 1. *Let S be a string of length $n \geq 1$ over the alphabet $\{a, b, c\}$ and let A, B, C respectively be the number of a's, b's and c's in S. Then:*

1. *S is a member of a* Crochemore Set *with generator a, if and only if:*

    - *S contains at least two distinct elements of $\{a, b, c\}$ and*
    - *for n even, A is even and (B,C) are odd; for n odd, A is odd and (B, C) are even.*

2. *Moreoever, if S_n is the number of distinct strings of length n over $\{a, b, c\}$, each of which is a member of some Crochemore Set with generator a, then:*

    - *for n even, $S_n = (3^n - 1)/4$*
    - *for n odd, $S_n = (3^n - 3)/4$*

Proof. *Part 1.* It is easy to prove, from the inductive definition of a CS, that Part 1 holds for every member of a CS. Conversely, if S has the properties of Part 1, then it is easy to see that there must exist at least one sequence of digits that (by a process which is the reverse to that for generating a string in a CS) reduces S to the single character a.

Part 2. It is easy to prove that of the 3^n distinct strings of length n over $\{a, b, c\}$, $(3^n + 1)/2$ have A even and $(3^n - 1)/2$ have A odd. For n even, the $(3^n + 1)/2$ strings with even A includes the strings consisting only of a's. This is disallowed by Part 1, leaving $(3^{n+1})/2 - 1 = (3^{n-1})/2$ strings. Each of these contains at least one character that is not a, and each has either (B, C) odd or (B, C) even. We can form pairs of strings such that one string of a pair differs from the other in just one respect. The first character that is not an a is a b in one and is a c in the other. Clearly *every* one of the $(3^{n-1})/2$ strings can be accounted for in this way. It follows that for n even, there are $(3^{n-1})/4$ strings with A even, (B, C) odd and containing a least two characters from $\{a, b, c\}$. For n odd, the proof is similar but we start with $(3^{n-1})/2$ strings with A odd. ∎

THEOREM 2. *The number of distinct Crochemore Sets of degree n is given by the $(n-1)$th Catalan number $(2(n-1))!/((n!(n-1)!))$.*

Proof. As we illustrate in Section 3, the replacement of one character in a string by a pair in an inductive construction of any string of a *Crochemore Set* is just the application of a *production* of a *regular language* that will generate all strings in that set. The *parse tree* of the string will have the *generator* of the set labelling its root. Its children will be labelled by the pair of characters that replaces the root, and so on. Notice the following interesting fact. The topology of the parse tree is determined uniquely by the sequence of digits D that parameterises the *Crochemore Set* $CS(G, D)$. Since D is employed for the inductive construction of every string in $CS(G, D)$, it follows that the parse tree for every element of a given Crochemore Set would have the *same topology* and this topology is unique for that set. It is also obvious that, uniquely, there is a Crochemore Set of degree n corresponding to every binary tree with n leaves. Thus, the number of Crochemore Sets of degree n is precisely the same as the number of binary trees with n leaves. As is well known, the nth *Catalan number* is the number of binary trees with $(n+1)$ leaves. ∎

As the proof of Theorem 2 hints, it is not difficult to generate a different picture of *Crochemore Sets* through the notions of regular languages and finite automata. This we do in the following section.

3   Finite automata, regular languages and Crochemore Sets

Here, $L(CS(G, D))$ and $G(CS(G, D))$ refer, respectively, to $CS(G, D)$ as a *language* and to the *grammar* generating that language. We are dealing with *regular languages* and our grammars will be presented in a well-known

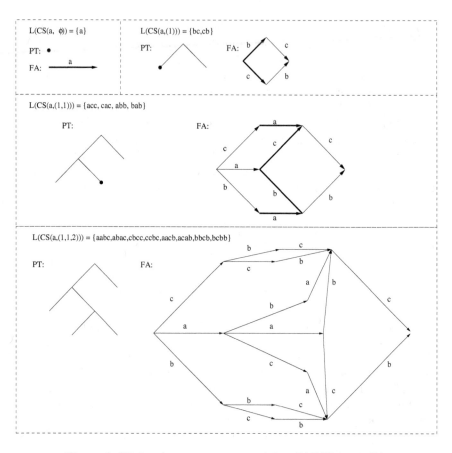

Figure 1. Finite Automaton recognising $L(CS(a, 1, 1, 2))$.

standard form for these in which each *production* replaces a single *non-terminal* by a *string of two non-terminals* or simply by a *single terminal*. As is well-known, every regular language has a recognising finite automaton and we describe the construction of these for any $L(CS(G, D))$. Every sentence of $L(CS(G, D))$, as was implied in the proof of Theorem 2, has a parse tree whose topology is the same for all sentences of $CS(G, D)$. Grammars, parse trees and finite automata here are all of a piece and we show, through one example, how they may, in general, all be developed together inductively for any given $L(CS(G, D))$.

Our example is for $L(CS(a, (1, 1, 2)))$. Inductively, we start with $L(CS(a, \varnothing))$ and build through $L(CS(a, (1)))$ and $L(CS(a, (1, 1)))$. The general step adds the next digit of D to the current digital sequence until the final goal is reached. Figure 1 concentrates on the parse trees (PT) and finite automaton (FA) in this process and for our example. Working through the sequence displayed, the inductive construction is clear. For $L(CS, (a, \varnothing))$, PT is a single vertex (which becomes the root of the subsequent PT's) and FA is a single edge which takes the machine from an initial state to the final state for an input a. We adopt the following conventions for the FA. The initial and final states are, respectively, the leftmost and rightmost states in the diagrams for each automaton. The automaton constructed are *deterministic* with the understanding that we do not show (for clarity) edges to a *dead* state which (using standard convention) is terminal for sentences not in the language. Thus, all the edges shown are on paths from the initial to final (accepting) state. From one step to the next, we essentially replace each edge in a subset (emboldened in the figures) by a *diamond* of edges. This is clearly illustrated in going from $L(CS(a, \varnothing))$ to $L(CS(a, (1)))$. In general, if the next digit of D to be added is d_i then the subset of edges involved is that set of edges at a distance d_i from the initial state. Where two or three edges in the subset emerge from the same state, then their replacement should involve some merging of the diamonds so as to retain determinacy of the automaton and this also minimises the number of states. This is illustrated in the second and thirds steps of the construction of Figure 1. A little thought easily reveals the rules for this. As far as the PT are concerned, at each step of the construction, we add a new pair of edges to the d_ith leaf from the left (emphasised by a heavy dot in the figures), now making this an internal vertex and creating two new leaves in the next PT.

Now consider the inductive construction of the grammar $G(CS(a, (1, 1, 2)))$. For all the following grammars, the *start symbol* is A_0. Also, for brevity, we say here once that the grammars appropriately include productions from the set $\{A \rightarrow a, B \rightarrow b, C \rightarrow c\}$ but we omit them explicitly below. Of course, a, b and c are the terminal symbols. We start with $G(CS(a, \varnothing))$ which has a single production $A_0 \rightarrow a$. In general, the following scheme is helpful. Bear in mind the parse tree for the target grammar $(G(CS(a, (1, 1, 2)))$ in this case) and uniquely associate a number with each internal vertex. Because there is an intimate relationship with normalised notation for the digital sequence D of a *Crochemore Set*, this would ideally be the depth first index (incremented by not counting leaves). Then each non-terminal of the grammar is uniquely associated with a particular internal vertex of the PT. This is indicated through its subscript. Several non-terminals are thus associated with any given internal node of PT, different associations arising from parsing different sentences. In the inductive step from one grammar to the next, we introduce new non-terminals associated with the newly created internal vertex of the PT and this is the key to understanding the construction. Thus, from $G(CS(a, \varnothing))$, we obtain $G(CS(a, (1)))$ which has productions $A_0 \rightarrow BC|CB$, then we form $G(CS(a, (1, 1)))$ with productions $A_0 \rightarrow B_1 C|C_1 B$, $B_1 \rightarrow AC|CA$ and $C_1 \rightarrow AB|BA$. Finally, we obtain $G(CS(a, (1, 1, 2)))$ with the production set:

$$
\begin{array}{lll}
A_0 \rightarrow B_1 C|C_1 B & B_1 \rightarrow AC_2|CA_2 & C_1 \rightarrow AB_2|BA_2 \\
A_2 \rightarrow BC|CB & B_2 \rightarrow AC|CA & C_2 \rightarrow AB|BA
\end{array}
$$

4   Colouring binary trees and Crochemore Sets

It will be clear that, through the grammatical generation of any string in a *Crochemore Set* (Section 3), we may associate a proper *3-vertex-colouring* of its binary parse tree. In such a colouring, the string is obtained by reading the colours of the leaves (from left to right). On the other hand, this string uniquely forces a proper 3-vertex colouring of the rest of the tree through an insistence that adjacent vertices have different colours. The string is a shorthand description of this 3-vertex-colouring of the tree. Therefore, an alternative view of a Crochemore Set is that is it is a shorthand specification of a certain subset of all proper 3-vertex-colourings of the binary tree. All possible proper 3-vertex-colourings of a binary tree are included in this subset *except* those in which any pair of sibling vertices are similarly coloured. The left-hand-side of Figure 2 is an illustration for the string $cabbc \in CS(a, (1, 1, 2, 2))$.

It now follows that we can obtain a proper *3-edge-colouring* of the tree by moving each individual colour from its vertex to the edge from its parent (we add an edge above the root for this construction). This movement of colour is illustrated in going from the left-hand-side of Figure 2 to the right-hand-side. So, again, we have another interpretation of *Crochemore Sets*. Each CS may be thought of as a shorthand specification of (in this case) *all possible* 3-edge-colourings of the binary tree in which G colours the edge above the root.

This association with edge-colouring of binary trees now allows us to make a connection between Crochemore Sets and the problem known as *Colouring Pairs of Binary Trees* (*CPBT*). Let (T_i, T_j) be *any* pair of binary trees, T_i having the same number of leaves as T_j. Then *CPBT* is to prove that there exists a string that, colouring the edges attached to leaves (from left-to-right) induces a proper 3-edge-colouring for *both* trees. That this is always possible is known only through the equivalence between *CPBT* and the famous *Four Colour Problem of Planar Maps* (*FCP*). This

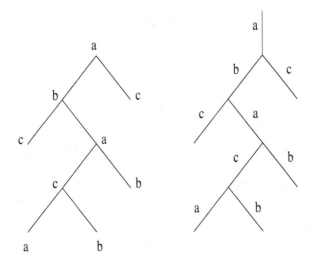

Figure 2. 3-vertex-colouring and 3-edge-colouring with *cabbc* $\in$ $CS(a, (1, 1, 2, 2))$.

equivalence is the subject, for example, of [1]. These observations now allow us to state the following theorem.

THEOREM 3. *For any pair of* Crochemore Sets, CS_i *and* CS_j *of the same degree and with the same generator,* $CS_i \cup CS_j \neq \varnothing$ *if and only if four colours are sufficient for a proper colouring of the regions of any planar map.*

The connection made by Theorem 3 is remarkable. It is famously known [4, 5, 6, 7] that four colours are sufficient to colour all planar maps. We have, therefore, the following corollary which would presumably be very difficult to prove otherwise.

COROLLARY 4. *Any two* Crochemore Sets *of the same degree and with the same generator have a* non-empty *intersection.*

5 Summary

We introduced the notion of *Crochemore Sets* and provided a number of related counting theorems. We also indicated their combinatorial richness through connections to *regular languages* and *colouring binary trees*. The non-trivial nature of questions related to *Crochemore Sets* is indicated through its connection to the notoriously difficult problem of *Four Colouring*

Planar Maps (*FCP*). Also note that, generally, questions of non-emptiness of regular language intersection are NP-hard; here we implicitly proved that intersections of languages formed by *Crochemore Sets* are non-empty and this is the only known proof of this fact.

We made the connection between the intersection of *Crochemore Sets* and *Colouring Pairs of Binary Trees* (*CPBT*). *CPBT* and its relationship with *FCP* has been the subject of a number of papers [1, 2, 3]. In [1], the authors showed that there is a linear time reduction between *FCP* and *CPBT*. A proof of the Corollary to Theorem 3 independent of the *FCT* would be of immense interest. There has long been a body of opinion that extant proofs [4, 5, 6, 7] of *FCT* are unsatisfactory, lacking conciseness and lucidity and requiring hours of electronic computation.

BIBLIOGRAPHY

[1] A. Czumaj and A.M. Gibbons, Guthrie's problem: new equivalences and rapid reductions, *Theoretical Computer Science*, Volume 154, Issue 1, 3-22, January 1996.

[2] Alan Gibbons and Paul Sant, Roration sequences and edge-colouring of binary tree pairs, *Theoretical Computer Science*, 326(1-3) pages 409-418, 2004.

[3] Alan Gibbons and Paul Sant, Stringology and The Four Colour Problem of Planar Maps, Strings Algorithmics, Volume 2 (*Texts in Algorithms*), C. Illiopoulos and T. Lecroq (Eds.), *King's College Publications*, 2004.

[4] K. Appel and W. Haken. Every planar map is four colourable. Part I. Discharging. *Illinois Journal of Mathematics* 21 (1977), 429-490.

[5] K. Appel, W. Haken and J. Koch. Every planar map is four colourable. Part II. Reducibility. *Illinois Journal of Mathematics* 21 (1977), 491-567.

[6] K. Appel and W. Haken. Every planar map is four colourable. *Contemporary Mathematics* 98 (1989) entire issue.

[7] N. Robertson, D.P. Sanders, P.D. Seymour and R. Thomas. A new proof of the four-colour theorem. Electron. Res. Announc. *Amer. Math. Soc.* 2 (1996), no. 1, 17-25 also: The four colour theorem, *J. Combin. Theory* Ser. B, 70 (1997), 2-44.

Alan Gibbons
Kings College London
Email: `alan.gibbons@kcl.ac.uk`

Paul Sant
University of Bedfordshire
Email: `Paul.Sant@beds.ac.uk`

Efficient Algorithms for Noise Propagation in Diffusion Tensor Imaging

IDA M. PU AND YUJI SHEN

ABSTRACT. We present efficient algorithms to determine the noise induced errors in magnetic resonance diffusion tensor imaging (MR-DTI). The error propagation theory is applied to establish a complete computational chain for the noise propagated from the acquired raw data to the calculated tensor parameters and to the various quantities derived from the tensor elements, including the eigenvalues, eigenvectors, mean diffusivity and typically used anisotropy indices. A simulation study is conducted based on the derived expressions and algorithms, and the results demonstrate the error behaviours associated with the signal to noise ratio in raw data.

1 Introduction

Magnetic resonance diffusion tensor imaging (MR-DTI) [1, 2, 3, 4] has been developed to become a quantitative technique for probing changes in tissue microstructures. Such changes have been found in various neurological diseases [5, 6, 7, 8, 9, 10, 11, 12, 13, 14], trauma [15], brain development [16] and ageing [17, 18]. The tissue microstructural integrity has therefore become an important issue in clinical applications, and MR-DTI technique becomes a particularly useful tool for neuroscience and for characterization of tissue structure [19].

It is water diffusion that enables tissue microstructure to be probed [4]. In MR-DTI, the directionally dependent water diffusion characteristics in tissues are characterised in terms of an effective diffusion tensor, which is a 3×3 symmetric matrix, and a second-order tensor in mathematical terms for each voxel.

The tensor contains important information about the tissue microstructure and fibre orientation. The eigenvalues of the diffusion tensor can be used to specify the intrinsic diffusivities in three orthogonal principal directions known as the principal eigenvectors. The principal eigenvector for the largest eigenvalue (λ_1) reflects the dominant local fibre tract orientation and is utilised by tractography, a technique that has been used to map the trajectory of the white matter tracts of the brain.

The rotationally invariant scalar quantities can also be derived from the diffusion tensor. The most commonly used scalar quantity is the mean diffusivity (D_{av}), which is used to measure the directionally averaged diffusivity, and the fractional anisotropy (FA) and relative anisotropy (RA), which characterise the directionally dependent anisotropic nature of the white matter tissue structure.

The measurement of a diffusion tensor for each voxel requires multiple image acquisitions with diffusion sensitivities in at least six non-collinear directions [20]. The six unknown tensor parameters at each voxel can then be determined either using the least-squares approach based on a multivariate linear regression technique [1], or using an analytical calculation method [20]. The eigenvalues and eigenvectors are determined consequently by diagonalization of the tensor matrix. The anisotropy indices, FA and RA, can be computed either from the calculated eigenvalues [3], or directly from the tensor invariants without tensor diagonalization [21], or directly from the raw data without decoding and diagonalization [22, 23].

Diffusion in the presence of magnetic field gradients generates MR signal attenuation, giving rise to a low signal-to-noise ratio (SNR) for the acquired diffusion-weighted (DW) images. The diffusion tensor calculated from noisy data inevitability suffers from accumulated errors, leading to unreliable tensor results. Noise in acquired DW images propagates in a chain first to the elements of the diffusion tensor, then to the tensor eigenvalues and eigenvectors and finally to the derived scalar parameters such as mean diffusivity, FA and RA. It is, therefore, important to know the precision of these derived results, and this has become an important issue for the use of DTI [3, 24, 25, 26, 27, 28, 29] and tractography [30, 31, 32]. Note that determination of the error level is a vital procedure in assessment of the reliability of a tensor result, and clinical studies rely on the precision of information to make correct judgements.

Analytical expressions and their computational efficiency have not, however, been fully established for noise propagation in DTI due to their complexity. Further theoretical investigation is necessary despite some valuable work previously. For example, Poonawalla et al [33] calculated the errors in FA and RA without treating the errors of the tensor eigenvalues directly. Thus the errors in those quantities that are solely derived from the tensor eigenvalues (e.g. the scalar indices for prolate and oblate tensors) cannot be determined by this approach. Anderson [34] used perturbation theory to investigate the noise effects on eigenvalue and eigenvector estimations and Chang et al [35] performed further validation. Unfortunately, the perturbation analysis is only valid for moderate and high SNR of DW images. This restricts its practical application since DW images are frequently acquired with low SNR. Similarly, Hasan et al [36] compared FA and RA analytically against errors without an explicit determination of the errors in the eigenvalues and/or eigenvectors. Recently, Koay et al [37] set up a framework for determining the underlying geometric relationship between the variability of a tensor-derived quantity and that of the DW signals. However, the methods that are used to determine the tensor elements in the above approaches are based on the least-squares methods, where the residuals cannot be zero in general. In addition, the matrix computation involved in these numerical approaches are expensive in terms of computational time and space. It is therefore necessary to explicitly determine the errors in the tensor elements and in the tensor eigenvalues and eigenvectors using fully analytical expressions in the assessment of the tensor result. One particular advantage of establishing such analytical expressions is to offer an efficient means in computation to determine tensor errors with minimum computational time.

Our contribution

In this study, we introduce a set of constant time algorithms that determine the errors in the diffusion tensor based on derived analytical expressions using analytical analysis and error propagation theory [38]. The work extends our preliminary results published in abstract form [39]. In this research, we focus on establishment of a complete set of analytical expressions for noise propagation in DTI and the corresponding efficient computational algorithms. The goal is to determine, in constant time, the errors in the diffusion tensor and the tensor derived parameters in terms of the underlying SNRs of the acquired DW images. The established analytical expressions include those of the errors propagated from the noise sampled to the tensor elements computed and those of the errors propagated further from the errors in the tensor elements to the tensor derived parameters, e.g., eigenvalues, eigenvectors, D_{av}, FA and RA. The established algorithms are based on the analytical expressions and offer an efficient computational means for accurate assessment of the reliability of a calculated tensor result. The derived algebraic expressions are useful for the decision making process in clinical DTI studies, providing a framework that can be used to model the required SNR (and/or number of acquisitions) for the detection of a given effect size in the tensor derived scalar metrics. In addition, the quantification of errors in the principal eigenvector has implications for tractography, where this information is utilized to map the white matter pathways of the brain.

2   Theory

The diffusion tensor is used to model three dimensional characteristics of diffusion within a tissue voxel. The diffusion MR signal S_i is often described as

$$S_i = S_0 e^{-b \mathbf{g}_i^T \mathbf{D} \mathbf{g}_i} \qquad \text{or} \qquad S_i = S_0 e^{-bADC_i} \qquad (1)$$

where S_0 is the baseline signal without diffusion gradients applied; b is the diffusion-weighting factor referred as b-value; $\mathbf{g}_i = (g_{x_i}, g_{y_i}, g_{z_i})$ is the direction of the i-th diffusion gradient; $\mathbf{D}$ is the second order diffusion tensor represented by a 3×3 symmetric matrix

$$\mathbf{D} \;=\; \begin{pmatrix} D_{xx} & D_{xy} & D_{xz} \\ D_{yx} & D_{yy} & D_{yz} \\ D_{zx} & D_{zy} & D_{zz} \end{pmatrix} \tag{2}$$

and ADC_i is apparent diffusion coefficient in i-th diffusion-weighting direction and is related to $\mathbf{D}$ by $ADC_i = \mathbf{g}_i^T \mathbf{D} \mathbf{g}_i$.

The diffusion tensor determined in Eq.(1) is an expression for an ideal situation where all the data acquired are free of noise. Subject to the noise in raw data, Eq.(1) should be replaced by

$$S_i' = S_0' e^{-b\mathbf{g}_i^T \mathbf{D}' \mathbf{g}_i} \qquad \text{or} \qquad S_i' = S_0' e^{-bADC_i'} \tag{3}$$

where the prime sign is used to indicate the measured signal with noise, or the tensor elements with errors.

Noise in raw data will introduce errors in computed tensor elements. When the tensor matrix is diagonalized to an eigensystem, these tensor errors are carried forward into the eigensystem of the diffusion tensor, leading to errors in tensor eigenvalues $(\lambda_1, \lambda_2, \lambda_3)$ and eigenvectors $(\hat{e}_1, \hat{e}_2, \hat{e}_3)$. When diffusion anisotropy indices or any other scalar parameters derived from the eigenvalues are calculated, the errors in the eigenvalues will further propagate to these indices and parameters. The process of error propagation can be simplified and summarized as follows, where σ represents the noise or potential error:

$$\left.\begin{array}{c} S_0 \pm \sigma_0 \\ S_1 \pm \sigma_1 \\ \vdots \\ S_6 \pm \sigma_6 \end{array}\right\} \Rightarrow \left.\begin{array}{c} ADC_1 \pm \sigma_{ADC_1} \\ ADC_2 \pm \sigma_{ADC_2} \\ \vdots \\ ADC_6 \pm \sigma_{ADC_6} \end{array}\right\} \Rightarrow$$

$$\left.\begin{array}{c} D_{xx} \pm \sigma_{xx} \\ D_{yy} \pm \sigma_{yy} \\ D_{zz} \pm \sigma_{zz} \\ D_{xy} \pm \sigma_{xy} \\ D_{yz} \pm \sigma_{yz} \\ D_{xz} \pm \sigma_{xz} \end{array}\right\} \Rightarrow \left.\begin{array}{c} \lambda_1 \pm \sigma_{\lambda_1} \\ \lambda_2 \pm \sigma_{\lambda_2} \\ \lambda_3 \pm \sigma_{\lambda_3} \\ \hat{e}_1 \pm \sigma_{\hat{e}_1} \\ \hat{e}_2 \pm \sigma_{\hat{e}_2} \\ \hat{e}_3 \pm \sigma_{\hat{e}_3} \end{array}\right\} \Rightarrow \begin{array}{c} RA \pm \sigma_{RA} \\ FA \pm \sigma_{FA} \\ C_A \pm \sigma_{C_A} \end{array}$$

In the remainder of this section, we first establish the mathematical expressions to precisely capture the errors in the diffusion tensor elements induced by noise in the raw data. We then determine the errors in the tensor eigenvalues and eigenvectors, and the errors in the diffusion anisotropy indices and in the mean diffusivity, using error propagation theory [38].

The error propagation formula for a general function of $f(x, y, z, \cdots)$ can be represented by

$$
\begin{aligned}
\sigma_f^2 &= \left(\frac{\partial f}{\partial x}\right)^2 \sigma_x^2 + \left(\frac{\partial f}{\partial y}\right)^2 \sigma_y^2 + \left(\frac{\partial f}{\partial z}\right)^2 \sigma_z^2 + \cdots \\
&\quad + 2\left(\frac{\partial f}{\partial x}\right)\left(\frac{\partial f}{\partial y}\right)\sigma_{xy}^2 + 2\left(\frac{\partial f}{\partial y}\right)\left(\frac{\partial f}{\partial z}\right)\sigma_{yz}^2 \\
&\quad + 2\left(\frac{\partial f}{\partial x}\right)\left(\frac{\partial f}{\partial z}\right)\sigma_{xz}^2 + \cdots
\end{aligned}
\tag{4}
$$

We assume all variables are uncorrelated as the covariance terms are much smaller than the variance terms as confirmed in DTI studies [33]. The error propagation formula can be simplified to

$$
\sigma_f^2 = \left(\frac{\partial f}{\partial x}\right)^2 \sigma_x^2 + \left(\frac{\partial f}{\partial y}\right)^2 \sigma_y^2 + \left(\frac{\partial f}{\partial z}\right)^2 \sigma_z^2 + \cdots
\tag{5}
$$

2.1 Determination of errors in diffusion tensor elements

We first demonstrate, in this section, how to determine six diffusion tensor elements analytically, and then derive expressions for the errors in these elements.

The relationship between tensor $\mathbf{D}$ and acquired ADC_i, $i = 1, \cdots 6$, in six non-colinear directions can be expressed as

$$
ADC_i = \frac{\ln S_0 - \ln S_i}{b} = \mathbf{g}_i^T \mathbf{D} \mathbf{g}_i
\tag{6}
$$

where $\mathbf{D}$ is represented by Eq.(2) and

$$
\mathbf{g}_i^T = (x_i, y_i, z_i)
$$

Eq.(6) can be further written as

$$
\begin{aligned}
ADC_i &= \begin{pmatrix} x_i & y_i & z_i \end{pmatrix} \begin{pmatrix} D_{xx} & D_{xy} & D_{xz} \\ D_{xy} & D_{yy} & D_{yz} \\ D_{xz} & D_{yz} & D_{zz} \end{pmatrix} \begin{pmatrix} x_i \\ y_i \\ z_i \end{pmatrix} \\
&= x_i^2 D_{xx} + y_i^2 D_{yy} + z_i^2 D_{zz} + \\
&\quad 2x_i y_i D_{xy} + 2y_i z_i D_{yz} + 2x_i z_i D_{xz}
\end{aligned}
\tag{7}
$$

In matrix form, it can be expressed as

$$
\begin{bmatrix} ADC_1 \\ ADC_2 \\ ADC_3 \\ ADC_4 \\ ADC_5 \\ ADC_6 \end{bmatrix} = \begin{bmatrix} x_1^2 & y_1^2 & z_1^2 & 2x_1y_1 & 2y_1z_1 & 2x_1z_1 \\ x_2^2 & y_2^2 & z_2^2 & 2x_2y_2 & 2y_2z_2 & 2x_2z_2 \\ x_3^2 & y_3^2 & z_3^2 & 2x_3y_3 & 2y_3z_3 & 2x_3z_3 \\ x_4^2 & y_4^2 & z_4^2 & 2x_4y_4 & 2y_4z_4 & 2x_4z_4 \\ x_5^2 & y_5^2 & z_5^2 & 2x_5y_5 & 2y_5z_5 & 2x_5z_5 \\ x_6^2 & y_6^2 & z_6^2 & 2x_6y_6 & 2y_6z_6 & 2x_6z_6 \end{bmatrix} \begin{bmatrix} D_{xx} \\ D_{yy} \\ D_{zz} \\ D_{xy} \\ D_{yz} \\ D_{xz} \end{bmatrix}
\tag{8}
$$

or simply written as

$$\mathbf{a} = \mathbf{Md}$$

Then $\mathbf{d}$ can be determined by

$$\mathbf{d} = \mathbf{M}^{-1}\mathbf{a} \tag{9}$$

where $\mathbf{a}$ is a vector of ADC; $\mathbf{d}$ is a vector of diffusion tensor elements, and $\mathbf{M}$ is the transfermation matrix, whose inverse can easily be determined. Hence for any diffusion encoding scheme with six non-colinear directions, an analytical solution always exists for determination of tensor elements.

Consider the following widely used six non-collinear diffusion-weighting directions: [1,0,1], [-1,0,1], [0,1,1], [0,1,-1], [1,1,0] and [-1,1,0]. The diffusion tensor elements can be analytically determined by Eq.(10)–(15):

$$D_{xz} = \frac{ADC_1 - ADC_2}{4} \tag{10}$$

$$D_{yz} = \frac{ADC_3 - ADC_4}{4} \tag{11}$$

$$D_{xy} = \frac{ADC_5 - ADC_6}{4} \tag{12}$$

$$D_{xx} = \frac{ADC_1 + ADC_2 - ADC_3 - ADC_4 + ADC_5 + ADC_6}{4} \tag{13}$$

$$D_{yy} = \frac{-ADC_1 - ADC_2 + ADC_3 + ADC_4 + ADC_5 + ADC_6}{4} \tag{14}$$

$$D_{zz} = \frac{ADC_1 + ADC_2 + ADC_3 + ADC_4 - ADC_5 - ADC_6}{4} \tag{15}$$

where

$$ADC_i = \frac{1}{b} \ln\left(\frac{S_0}{S_i}\right), \quad i = 1, \cdots, 6 \tag{16}$$

and S_i is the intensity value for the i-th DW signal and S_0 is the intensity value for the baseline signal acquired with $b = 0$ [20].

Consider the meausred signals with noise. Eq.(16) can be rewritten as

$$ADC'_i = \frac{1}{b} \ln\left(\frac{S'_0}{S'_i}\right), \quad i = 1, \cdots, 6 \tag{17}$$

where

$$S'_0 = S_0 + \sigma_0 \qquad \text{and} \qquad S'_i = S_i + \sigma_i$$

We use σ_0 to represent the noise for non-diffusion-weighted signal and σ_i the noise for the i-th diffusion weighted signal.

The error of ADC_i can be described as

$$\sigma_{ADC_i} = |ADC'_i - ADC_i| \tag{18}$$

where ADC'_i represents that with error and ADC_i without. Applying Eq.(16) and Eq.(17), we have

$$
\begin{aligned}
\sigma_{ADC_i} &= |ADC'_i - ADC_i| \\
&= \frac{1}{b} \left| \ln\left(\frac{S_0 + \sigma_0}{S_i + \sigma_i}\right) - \ln\left(\frac{S_0}{S_i}\right) \right| \\
&= \frac{1}{b} \left| \ln\left(\frac{1 + \frac{1}{SNR_0}}{1 + \frac{1}{SNR_i}}\right) \right|
\end{aligned}
\tag{19}
$$

where

$$SNR_0 = \frac{S_0}{\sigma_0} \qquad \text{and} \qquad SNR_i = \frac{S_i}{\sigma_i}$$

We now derive the expressions for the errors of diffusion tensor elements ($\sigma_{xz}, \sigma_{yz}, \sigma_{xy}, \sigma_{xx}, \sigma_{yy}$ and σ_{zz}) individually:

$$
\begin{aligned}
\sigma_{xz} &= |D'_{xz} - D_{xz}| \\
&= \left| \frac{(ADC_1 + \sigma_{ADC_1}) - (ADC_2 + \sigma_{ADC_2})}{4} - \frac{ADC_1 - ADC_2}{4} \right| \\
&= \left| \frac{\sigma_{ADC_1} - \sigma_{ADC_2}}{4} \right| \\
&= \frac{1}{4b} \left| \ln\left(\frac{1 + \frac{1}{SNR_0}}{1 + \frac{1}{SNR_1}}\right) - \ln\left(\frac{1 + \frac{1}{SNR_0}}{1 + \frac{1}{SNR_2}}\right) \right| \\
&= \frac{1}{4b} \left| \ln\left(\frac{1 + SNR_2^{-1}}{1 + SNR_1^{-1}}\right) \right|
\end{aligned}
\tag{20}
$$

Similarly,

$$
\begin{aligned}
\sigma_{yz} &= |D'_{yz} - D_{yz}| \\
&= \frac{1}{4b} \left| \ln\left(\frac{1 + SNR_4^{-1}}{1 + SNR_3^{-1}}\right) \right|
\end{aligned}
\tag{21}
$$

$$\sigma_{xy} = |D'_{xy} - D_{xy}|$$

$$= \frac{1}{4b}\left|\ln\left(\frac{1 + SNR_6^{-1}}{1 + SNR_5^{-1}}\right)\right| \tag{22}$$

$$\sigma_{xx} = |D'_{xx} - D_{xx}|$$

$$= \frac{1}{4b}\left|\ln\left(\frac{(1 + SNR_3^{-1})(1 + SNR_4^{-1})(1 + SNR_0^{-1})^2}{(1 + SNR_1^{-1})(1 + SNR_2^{-1})(1 + SNR_5^{-1})(1 + SNR_6^{-1})}\right)\right| \tag{23}$$

$$\sigma_{yy} = |D'_{yy} - D_{yy}|$$

$$= \frac{1}{4b}\left|\ln\left(\frac{(1 + SNR_1^{-1})(1 + SNR_2^{-1})(1 + SNR_0^{-1})^2}{(1 + SNR_3^{-1})(1 + SNR_4^{-1})(1 + SNR_5^{-1})(1 + SNR_6^{-1})}\right)\right| \tag{24}$$

$$\sigma_{zz} = |D'_{zz} - D_{zz}|$$

$$= \frac{1}{4b}\left|\ln\left(\frac{(1 + SNR_5^{-1})(1 + SNR_6^{-1})(1 + SNR_0^{-1})^2}{(1 + SNR_1^{-1})(1 + SNR_2^{-1})(1 + SNR_3^{-1})(1 + SNR_4^{-1})}\right)\right| \tag{25}$$

2.2 Determination of errors in tensor eigenvalues

The eigenvalues of the diffusion tensor can be analytically determined by (see [40] and [41])

$$\lambda_1 = \frac{I_1}{3} + \frac{2}{3}\sqrt{I_1^2 - 3I_2}\cos(\phi) \tag{26}$$

$$\lambda_2 = \frac{I_1}{3} + \frac{2}{3}\sqrt{I_1^2 - 3I_2}\cos(\phi - \frac{2}{3}\pi) \tag{27}$$

$$\lambda_3 = \frac{I_1}{3} + \frac{2}{3}\sqrt{I_1^2 - 3I_2}\cos(\phi + \frac{2}{3}\pi) \tag{28}$$

where I_1, I_2, I_3 are tensor invariants and can be expressed respectively by

$$I_1 = D_{xx} + D_{yy} + D_{zz} \tag{29}$$

$$I_2 = D_{xx}D_{yy} + D_{xx}D_{zz} + D_{yy}D_{zz} - (D_{xy}^2 + D_{yz}^2 + D_{xz}^2) \tag{30}$$

$$I_3 = D_{xx}D_{yy}D_{zz} + 2D_{xy}D_{yz}D_{xz} - (D_{zz}D_{xy}^2 + D_{yy}D_{xz}^2 + D_{xx}D_{yz}^2) \quad (31)$$

and

$$\phi = \frac{1}{3}arccos(v) \tag{32}$$

where

$$v = \frac{2I_1^3 - 9I_1I_2 + 27I_3}{2(I_1^2 - 3I_2)^{3/2}} \tag{33}$$

Using propagation of errors to Eqs.(26)–(28), the errors in eigenvalues can be calculated by

$$\sigma_{\lambda_1}^2 = \left(\frac{\partial \lambda_1}{\partial I_1}\right)^2 \sigma_{I_1}^2 + \left(\frac{\partial \lambda_1}{\partial I_2}\right)^2 \sigma_{I_2}^2 + \left(\frac{\partial \lambda_1}{\partial I_3}\right)^2 \sigma_{I_3}^2 \tag{34}$$

$$\sigma_{\lambda_2}^2 = \left(\frac{\partial \lambda_2}{\partial I_1}\right)^2 \sigma_{I_1}^2 + \left(\frac{\partial \lambda_2}{\partial I_2}\right)^2 \sigma_{I_2}^2 + \left(\frac{\partial \lambda_2}{\partial I_3}\right)^2 \sigma_{I_3}^2 \tag{35}$$

$$\sigma_{\lambda_3}^2 = \left(\frac{\partial \lambda_3}{\partial I_1}\right)^2 \sigma_{I_1}^2 + \left(\frac{\partial \lambda_3}{\partial I_2}\right)^2 \sigma_{I_2}^2 + \left(\frac{\partial \lambda_3}{\partial I_3}\right)^2 \sigma_{I_3}^2 \tag{36}$$

Conducting a calculation of error propagation to the three tensor invariant expressions (Eqs.(29)–(31)), we have

$$\begin{aligned}
\sigma_{I_1}^2 &= \left(\frac{\partial I_1}{\partial D_{xx}}\right)^2 \sigma_{xx}^2 + \left(\frac{\partial I_1}{\partial D_{yy}}\right)^2 \sigma_{yy}^2 + \left(\frac{\partial I_1}{\partial D_{zz}}\right)^2 \sigma_{zz}^2 \\
&= \sigma_{xx}^2 + \sigma_{yy}^2 + \sigma_{zz}^2
\end{aligned} \tag{37}$$

$$\begin{aligned}
\sigma_{I_2}^2 &= \left(\frac{\partial I_2}{\partial D_{xx}}\right)^2 \sigma_{xx}^2 + \left(\frac{\partial I_2}{\partial D_{yy}}\right)^2 \sigma_{yy}^2 + \left(\frac{\partial I_2}{\partial D_{zz}}\right)^2 \sigma_{zz}^2 \\
&+ \left(\frac{\partial I_2}{\partial D_{xy}}\right)^2 \sigma_{xy}^2 + \left(\frac{\partial I_2}{\partial D_{xz}}\right)^2 \sigma_{xz}^2 + \left(\frac{\partial I_2}{\partial D_{yz}}\right)^2 \sigma_{yz}^2 \\
&= (D_{yy} + D_{zz})^2\sigma_{xx}^2 + (D_{xx} + D_{zz})^2\sigma_{yy}^2 + (D_{xx} + D_{yy})^2\sigma_{zz}^2 \\
&+ 4(D_{xy}^2\sigma_{xy}^2 + D_{xz}^2\sigma_{xz}^2 + D_{yz}^2\sigma_{yz}^2)
\end{aligned} \tag{38}$$

$$
\begin{aligned}
\sigma_{I_3}^2 &= \left(\frac{\partial I_3}{\partial D_{xx}}\right)^2 \sigma_{xx}^2 + \left(\frac{\partial I_3}{\partial D_{yy}}\right)^2 \sigma_{yy}^2 + \left(\frac{\partial I_3}{\partial D_{zz}}\right)^2 \sigma_{zz}^2 \\
&+ \left(\frac{\partial I_3}{\partial D_{xy}}\right)^2 \sigma_{xy}^2 + \left(\frac{\partial I_3}{\partial D_{xz}}\right)^2 \sigma_{xz}^2 + \left(\frac{\partial I_3}{\partial D_{yz}}\right)^2 \sigma_{yz}^2 \\
&= (D_{yy}D_{zz} - D_{yz}^2)^2\sigma_{xx}^2 + (D_{xx}D_{zz} - D_{xz}^2)^2\sigma_{yy}^2 + (D_{xx}D_{yy} - D_{xy}^2)^2\sigma_{zz}^2 \\
&+ 4(D_{xz}D_{yz} - D_{zz}D_{xy})^2\sigma_{xy}^2 + 4(D_{xy}D_{yz} - D_{yy}D_{xz})^2\sigma_{xz}^2 \\
&+ 4(D_{xy}D_{xz} - D_{xx}D_{yz})^2\sigma_{yz}^2
\end{aligned}
\tag{39}
$$

The partial derivatives of the eigenvalue expressions (Eqs.(26) –(28)) are calculated to be

$$
\frac{\partial \lambda_1}{\partial I_1} = \frac{1}{3} + \frac{2I_1}{3\sqrt{I_1^2 - 3I_2}}cos(\phi) - \frac{2}{9}\sqrt{I_1^2 - 3I_2}sin(\phi)\frac{\partial(arccos(v))}{\partial I_1}
\tag{40}
$$

$$
\frac{\partial \lambda_1}{\partial I_2} = -\frac{cos(\phi)}{\sqrt{I_1^2 - 3I_2}} - \frac{2}{9}\sqrt{I_1^2 - 3I_2}sin(\phi)\frac{\partial(arccos(v))}{\partial I_2}
\tag{41}
$$

$$
\frac{\partial \lambda_1}{\partial I_3} = -\frac{2}{9}\sqrt{I_1^2 - 3I_2}sin(\phi)\frac{\partial(arccos(v))}{\partial I_3}
\tag{42}
$$

$$
\begin{aligned}
\frac{\partial \lambda_2}{\partial I_1} &= \frac{1}{3} + \frac{2I_1}{3\sqrt{I_1^2 - 3I_2}}cos\left(\phi - \frac{2}{3}\pi\right) \\
&- \frac{2}{9}\sqrt{I_1^2 - 3I_2}sin\left(\phi - \frac{2}{3}\pi\right)\frac{\partial(arccos(v))}{\partial I_1}
\end{aligned}
\tag{43}
$$

$$
\begin{aligned}
\frac{\partial \lambda_2}{\partial I_2} &= -\frac{1}{\sqrt{I_1^2 - 3I_2}}cos\left(\phi - \frac{2}{3}\pi\right) \\
&- \frac{2}{9}\sqrt{I_1^2 - 3I_2}sin\left(\phi - \frac{2}{3}\pi\right)\frac{\partial(arccos(v))}{\partial I_2}
\end{aligned}
\tag{44}
$$

$$
\frac{\partial \lambda_2}{\partial I_3} = -\frac{2}{9}\sqrt{I_1^2 - 3I_2}sin\left(\phi - \frac{2}{3}\pi\right)\frac{\partial(arccos(v))}{\partial I_3}
\tag{45}
$$

$$
\begin{aligned}
\frac{\partial \lambda_3}{\partial I_1} &= \frac{1}{3} + \frac{2I_1}{3\sqrt{I_1^2 - 3I_2}}cos\left(\phi + \frac{2}{3}\pi\right) \\
&- \frac{2}{9}\sqrt{I_1^2 - 3I_2}sin\left(\phi + \frac{2}{3}\pi\right)\frac{\partial(arccos(v))}{\partial I_1}
\end{aligned}
\tag{46}
$$

$$\frac{\partial \lambda_3}{\partial I_2} = -\frac{1}{\sqrt{I_1^2 - 3I_2}} \cos\left(\phi + \frac{2}{3}\pi\right)$$
$$- \frac{2}{9}\sqrt{I_1^2 - 3I_2} \sin\left(\phi + \frac{2}{3}\pi\right) \frac{\partial(arccos(v))}{\partial I_2} \tag{47}$$

$$\frac{\partial \lambda_3}{\partial I_3} = -\frac{2}{9}\sqrt{I_1^2 - 3I_2} \sin\left(\phi + \frac{2}{3}\pi\right) \frac{\partial(arccos(v))}{\partial I_3} \tag{48}$$

where

$$\frac{\partial(arccos(v))}{\partial I_1} = -\frac{1}{\sqrt{1 - v^2}} \frac{\partial v}{\partial I_1} \tag{49}$$

$$\frac{\partial(arccos(v))}{\partial I_2} = -\frac{1}{\sqrt{1 - v^2}} \frac{\partial v}{\partial I_2} \tag{50}$$

$$\frac{\partial(arccos(v))}{\partial I_3} = -\frac{1}{\sqrt{1 - v^2}} \frac{\partial v}{\partial I_3} \tag{51}$$

and

$$\frac{\partial v}{\partial I_1} = \frac{3}{2}(2I_1^2 - 3I_2)(I_1^2 - 3I_2)^{-3/2}$$
$$- \frac{3}{2}I_1(2I_1^3 - 9I_1 I_2 + 27I_3)(I_1^2 - 3I_2)^{-5/2} \tag{52}$$

$$\frac{\partial v}{\partial I_2} = -\frac{9}{2}I_1(I_1^2 - 3I_2)^{-3/2} + \frac{9}{4}(2I_1^3 - 9I_1 I_2 + 27I_3)(I_1^2 - 3I_2)^{-5/2} \tag{53}$$

$$\frac{\partial v}{\partial I_3} = \frac{27}{2}(I_1^2 - 3I_2)^{-3/2} \tag{54}$$

Substituting Eqs.(37)–(48) into Eqs.(34)–(36) and using the errors in diffusion tensor elements expressed by Eqs.(20)–(25), the errors in eigenvalues can be analytically determined.

2.3　Determination of errors in tensor eigenvectors

The tensor eigenvectors can also be analytically determined. For the i-th eigenvalue λ_i ($i = 1, 2, 3$), the corresponding eigenvector $\hat{e}_i$ can be calculated by (see Ref.[40])

$$e_{x_i} = \frac{U_1 U_2}{\sqrt{U_1^2 U_2^2 + U_2^2 U_3^2 + U_1^2 U_3^2}} \tag{55}$$

$$e_{y_i} = \frac{U_2 U_3}{\sqrt{U_1^2 U_2^2 + U_2^2 U_3^2 + U_1^2 U_3^2}} \tag{56}$$

$$e_{z_i} = \frac{U_1 U_3}{\sqrt{U_1^2 U_2^2 + U_2^2 U_3^2 + U_1^2 U_3^2}} \tag{57}$$

where

$$U_1 = D_{xy} D_{yz} - (D_{yy} - \lambda_i) D_{xz} \tag{58}$$

$$U_2 = D_{xz} D_{yz} - (D_{zz} - \lambda_i) D_{xy} \tag{59}$$

$$U_3 = D_{xz} D_{xy} - (D_{xx} - \lambda_i) D_{yz} \tag{60}$$

Applying propagation of errors to the eigenvector calculation equations (Eqs.(55)–(57)), the errors in eigenvectors can be expressed by

$$\sigma_{e_{x_i}}^2 = \left(\frac{\partial e_{x_i}}{\partial U_1}\right)^2 \sigma_{U_1}^2 + \left(\frac{\partial e_{x_i}}{\partial U_2}\right)^2 \sigma_{U_2}^2 + \left(\frac{\partial e_{x_i}}{\partial U_3}\right)^2 \sigma_{U_3}^2 \tag{61}$$

$$\sigma_{e_{y_i}}^2 = \left(\frac{\partial e_{y_i}}{\partial U_1}\right)^2 \sigma_{U_1}^2 + \left(\frac{\partial e_{y_i}}{\partial U_2}\right)^2 \sigma_{U_2}^2 + \left(\frac{\partial e_{y_i}}{\partial U_3}\right)^2 \sigma_{U_3}^2 \tag{62}$$

$$\sigma_{e_{z_i}}^2 = \left(\frac{\partial e_{z_i}}{\partial U_1}\right)^2 \sigma_{U_1}^2 + \left(\frac{\partial e_{z_i}}{\partial U_2}\right)^2 \sigma_{U_2}^2 + \left(\frac{\partial e_{z_i}}{\partial U_3}\right)^2 \sigma_{U_3}^2 \tag{63}$$

Applying error propagation to the terms of U_1, U_2 and U_3 represented by Eqs.(58)–(60) individually, we have

$$\sigma_{U_1}^2 = D_{yz}^2 \sigma_{xy}^2 + D_{xy}^2 \sigma_{yz}^2 + D_{xz}^2 \sigma_{yy}^2 + D_{xz}^2 \sigma_{\lambda_i}^2 + (\lambda_i - D_{yy})^2 \sigma_{xz}^2 \tag{64}$$

$$\sigma_{U_2}^2 = D_{yz}^2 \sigma_{xz}^2 + D_{xz}^2 \sigma_{yz}^2 + D_{xy}^2 \sigma_{zz}^2 + D_{xy}^2 \sigma_{\lambda_i}^2 + (\lambda_i - D_{zz})^2 \sigma_{xy}^2 \tag{65}$$

$$\sigma_{U_3}^2 = D_{xy}^2 \sigma_{xz}^2 + D_{xz}^2 \sigma_{xy}^2 + D_{yz}^2 \sigma_{xx}^2 + D_{yz}^2 \sigma_{\lambda_i}^2 + (\lambda_i - D_{xx})^2 \sigma_{yz}^2 \tag{66}$$

The partial derivatives of the eigenvector expressions (Eqs.(55)–(57)) are given by

$$\frac{\partial e_{x_i}}{\partial U_1} = \frac{U_2}{(U_1^2 U_2^2 + U_2^2 U_3^2 + U_1^2 U_3^2)^{1/2}} - \frac{U_1^2 U_2 (U_2^2 + U_3^2)}{(U_1^2 U_2^2 + U_2^2 U_3^2 + U_1^2 U_3^2)^{3/2}} \tag{67}$$

$$\frac{\partial e_{x_i}}{\partial U_2} = \frac{U_1}{(U_1^2 U_2^2 + U_2^2 U_3^2 + U_1^2 U_3^2)^{1/2}} - \frac{U_2^2 U_1 (U_1^2 + U_3^2)}{(U_1^2 U_2^2 + U_2^2 U_3^2 + U_1^2 U_3^2)^{3/2}} \quad (68)$$

$$\frac{\partial e_{x_i}}{\partial U_3} = -\frac{U_1 U_2 U_3 (U_1^2 + U_2^2)}{(U_1^2 U_2^2 + U_2^2 U_3^2 + U_1^2 U_3^2)^{3/2}} \quad (69)$$

$$\frac{\partial e_{y_i}}{\partial U_1} = -\frac{U_1 U_2 U_3 (U_2^2 + U_3^2)}{(U_1^2 U_2^2 + U_2^2 U_3^2 + U_1^2 U_3^2)^{3/2}} \quad (70)$$

$$\frac{\partial e_{y_i}}{\partial U_2} = \frac{U_3}{(U_1^2 U_2^2 + U_2^2 U_3^2 + U_1^2 U_3^2)^{1/2}} - \frac{U_2^2 U_3 (U_1^2 + U_3^2)}{(U_1^2 U_2^2 + U_2^2 U_3^2 + U_1^2 U_3^2)^{3/2}} \quad (71)$$

$$\frac{\partial e_{y_i}}{\partial U_3} = \frac{U_2}{(U_1^2 U_2^2 + U_2^2 U_3^2 + U_1^2 U_3^2)^{1/2}} - \frac{U_3^2 U_2 (U_1^2 + U_2^2)}{(U_1^2 U_2^2 + U_2^2 U_3^2 + U_1^2 U_3^2)^{3/2}} \quad (72)$$

$$\frac{\partial e_{z_i}}{\partial U_1} = \frac{U_3}{(U_1^2 U_2^2 + U_2^2 U_3^2 + U_1^2 U_3^2)^{1/2}} - \frac{U_1^2 U_3 (U_2^2 + U_3^2)}{(U_1^2 U_2^2 + U_2^2 U_3^2 + U_1^2 U_3^2)^{3/2}} \quad (73)$$

$$\frac{\partial e_{z_i}}{\partial U_2} = -\frac{U_1 U_2 U_3 (U_1^2 + U_2^2)}{(U_1^2 U_2^2 + U_2^2 U_3^2 + U_1^2 U_3^2)^{3/2}} \quad (74)$$

$$\frac{\partial e_{z_i}}{\partial U_3} = \frac{U_1}{(U_1^2 U_2^2 + U_2^2 U_3^2 + U_1^2 U_3^2)^{1/2}} - \frac{U_3^2 U_1 (U_1^2 + U_2^2)}{(U_1^2 U_2^2 + U_2^2 U_3^2 + U_1^2 U_3^2)^{3/2}} \quad (75)$$

Putting Eqs.(64)–(75) into Eqs.(61)–(63) and using the errors in diffusion tensor elements expressed by Eqs.(20)–(25) and the errors in tensor eigenvalues represented by Eqs.(34)–(36), the errors in eigenvectors can be analytically determined.

2.4   Determination of errors in diffusion anisotropy indices

FA and RA [3] are the two most commonly used diffusion anisotropy indices for the scalar measurement of diffusion anisotropy. The other commonly used diffusion anisotropy index (C_A) is defined by Westin et al [42]. Here we only discuss and calculate the errors in these three indices because other published indices are rarely used. The approach described here, however, can be extended to determine the errors in any diffusion indices derived from the tensor.

The FA, RA and C_A are defined respectively as

$$FA = \frac{1}{\sqrt{2}} \frac{\sqrt{(\lambda_1 - \lambda_2)^2 + (\lambda_2 - \lambda_3)^2 + (\lambda_3 - \lambda_1)^2}}{\sqrt{\lambda_1^2 + \lambda_2^2 + \lambda_3^2}} \tag{76}$$

$$RA = \frac{1}{\sqrt{2}} \frac{\sqrt{(\lambda_1 - \lambda_2)^2 + (\lambda_2 - \lambda_3)^2 + (\lambda_3 - \lambda_1)^2}}{\lambda_1 + \lambda_2 + \lambda_3} \tag{77}$$

$$C_A = 1 - \frac{3\lambda_3}{\lambda_1 + \lambda_2 + \lambda_3} \tag{78}$$

Applying propagation of errors to FA, we have

$$\sigma_{FA}^2 = \left(\frac{\partial FA}{\partial \lambda_1}\right)^2 \sigma_{\lambda_1}^2 + \left(\frac{\partial FA}{\partial \lambda_2}\right)^2 \sigma_{\lambda_2}^2 + \left(\frac{\partial FA}{\partial \lambda_3}\right)^2 \sigma_{\lambda_3}^2 \tag{79}$$

The partial derivatives of FA to λ_1, λ_2 and λ_3 are given below

$$\frac{\partial FA}{\partial \lambda_1} = \frac{1}{\sqrt{2}} \left(\frac{2\lambda_1 - \lambda_2 - \lambda_3}{\sqrt{\lambda_1^2 + \lambda_2^2 + \lambda_3^2}\sqrt{(\lambda_1 - \lambda_2)^2 + (\lambda_2 - \lambda_3)^2 + (\lambda_3 - \lambda_1)^2}} \right.$$
$$\left. - \frac{\lambda_1 \sqrt{(\lambda_1 - \lambda_2)^2 + (\lambda_2 - \lambda_3)^2 + (\lambda_3 - \lambda_1)^2}}{(\lambda_1^2 + \lambda_2^2 + \lambda_3^2)^{3/2}} \right) \tag{80}$$

$$\frac{\partial FA}{\partial \lambda_2} = \frac{1}{\sqrt{2}} \left(\frac{2\lambda_2 - \lambda_3 - \lambda_1}{\sqrt{\lambda_1^2 + \lambda_2^2 + \lambda_3^2}\sqrt{(\lambda_1 - \lambda_2)^2 + (\lambda_2 - \lambda_3)^2 + (\lambda_3 - \lambda_1)^2}} \right.$$
$$\left. - \frac{\lambda_2 \sqrt{(\lambda_1 - \lambda_2)^2 + (\lambda_2 - \lambda_3)^2 + (\lambda_3 - \lambda_1)^2}}{(\lambda_1^2 + \lambda_2^2 + \lambda_3^2)^{3/2}} \right) \tag{81}$$

$$\frac{\partial FA}{\partial \lambda_3} = \frac{1}{\sqrt{2}} \left(\frac{2\lambda_3 - \lambda_2 - \lambda_1}{\sqrt{\lambda_1^2 + \lambda_2^2 + \lambda_3^2}\sqrt{(\lambda_1 - \lambda_2)^2 + (\lambda_2 - \lambda_3)^2 + (\lambda_3 - \lambda_1)^2}} \right.$$
$$\left. - \frac{\lambda_3 \sqrt{(\lambda_1 - \lambda_2)^2 + (\lambda_2 - \lambda_3)^2 + (\lambda_3 - \lambda_1)^2}}{(\lambda_1^2 + \lambda_2^2 + \lambda_3^2)^{3/2}} \right) \tag{82}$$

Substituting Eqs.(80)–(82) into Eq.(79) and using the errors of eigenvalues defined by Eqs.(34)–(36), the error in FA can be calculated analytically.

Similarly, we apply propagation of errors to RA and have

$$\sigma^2_{RA} = \left(\frac{\partial RA}{\partial \lambda_1}\right)^2 \sigma^2_{\lambda_1} + \left(\frac{\partial RA}{\partial \lambda_2}\right)^2 \sigma^2_{\lambda_2} + \left(\frac{\partial RA}{\partial \lambda_3}\right)^2 \sigma^2_{\lambda_3} \tag{83}$$

The partial derivatives of RA to λ_1, λ_2 and λ_3 are calculated to be

$$\frac{\partial RA}{\partial \lambda_1} = \frac{1}{\sqrt{2}} \left[\frac{2\lambda_1 - \lambda_2 - \lambda_3}{(\lambda_1 + \lambda_2 + \lambda_3)\sqrt{(\lambda_1 - \lambda_2)^2 + (\lambda_2 - \lambda_3)^2 + (\lambda_3 - \lambda_1)^2}} \right.$$
$$\left. - \frac{\sqrt{(\lambda_1 - \lambda_2)^2 + (\lambda_2 - \lambda_3)^2 + (\lambda_3 - \lambda_1)^2}}{(\lambda_1 + \lambda_2 + \lambda_3)^2} \right] \tag{84}$$

$$\frac{\partial RA}{\partial \lambda_2} = \frac{1}{\sqrt{2}} \left[\frac{2\lambda_2 - \lambda_1 - \lambda_3}{(\lambda_1 + \lambda_2 + \lambda_3)\sqrt{(\lambda_1 - \lambda_2)^2 + (\lambda_2 - \lambda_3)^2 + (\lambda_3 - \lambda_1)^2}} \right.$$
$$\left. - \frac{\sqrt{(\lambda_1 - \lambda_2)^2 + (\lambda_2 - \lambda_3)^2 + (\lambda_3 - \lambda_1)^2}}{(\lambda_1 + \lambda_2 + \lambda_3)^2} \right] \tag{85}$$

$$\frac{\partial RA}{\partial \lambda_3} = \frac{1}{\sqrt{2}} \left[\frac{2\lambda_3 - \lambda_2 - \lambda_1}{(\lambda_1 + \lambda_2 + \lambda_3)\sqrt{(\lambda_1 - \lambda_2)^2 + (\lambda_2 - \lambda_3)^2 + (\lambda_3 - \lambda_1)^2}} \right.$$
$$\left. - \frac{\sqrt{(\lambda_1 - \lambda_2)^2 + (\lambda_2 - \lambda_3)^2 + (\lambda_3 - \lambda_1)^2}}{(\lambda_1 + \lambda_2 + \lambda_3)^2} \right] \tag{86}$$

Substitution of Eqs.(84)–(86) into Eq.(83) and use of the errors of eigenvalues determined by Eqs.(34)–(36), the error in RA can be analytically determined.

Again, we use propagation of errors to C_A and have

$$\sigma^2_{C_A} = \left(\frac{\partial C_A}{\partial \lambda_1}\right)^2 \sigma^2_{\lambda_1} + \left(\frac{\partial C_A}{\partial \lambda_2}\right)^2 \sigma^2_{\lambda_2} + \left(\frac{\partial C_A}{\partial \lambda_3}\right)^2 \sigma^2_{\lambda_3} \tag{87}$$

The partial derivatives of C_A to λ_1, λ_2 and λ_3 are

$$\frac{\partial C_A}{\partial \lambda_1} = \frac{3\lambda_3}{(\lambda_1 + \lambda_2 + \lambda_3)^2} \tag{88}$$

$$\frac{\partial C_A}{\partial \lambda_2} = \frac{3\lambda_3}{(\lambda_1 + \lambda_2 + \lambda_3)^2} \qquad (89)$$

$$\frac{\partial C_A}{\partial \lambda_3} = \frac{-3\,(\lambda_1 + \lambda_2)}{(\lambda_1 + \lambda_2 + \lambda_3)^2} \qquad (90)$$

Replacing Eqs.(88)–(90) back into Eq.(87) and using the errors of eigenvalues quantified by Eqs.(34)–(36) , the error in C_A can also be determined analytically.

Note that, since $\sigma_{FA} = 1/3(FA/RA)^3\sigma_{RA}$ [36], only one error (σ_{FA} or σ_{RA}) needs to be calculated, and the other one can be derived from the formula.

2.5   Determination of errors in mean diffusivity

The mean diffusivity is defined as

$$D_{av} = \frac{1}{3}(D_{xx} + D_{yy} + D_{zz}) \qquad (91)$$

With application of error propagation to the expression of D_{av}, we have

$$
\begin{aligned}
\sigma_{D_{av}}^2 &= \left(\frac{\partial D_{av}}{\partial D_{xx}}\right)^2 \sigma_{xx}^2 + \left(\frac{\partial D_{av}}{\partial D_{yy}}\right)^2 \sigma_{yy}^2 + \left(\frac{\partial D_{av}}{\partial D_{zz}}\right)^2 \sigma_{zz}^2 \\
&= \frac{1}{9}(\sigma_{xx}^2 + \sigma_{yy}^2 + \sigma_{zz}^2)
\end{aligned}
\qquad (92)
$$

The error of mean diffusivity depends on the errors in the tensor diagonal elements.

3   Algorithms

We derive eight algorithms for various stages of the error propergation chain based on the theories discussed in the previous sections.

Algorithm 1 computes the tensor elements. Algorithm 2 computes the tensor invariants. Algorithm 3 computes the eigenvalues. Algorithm 4 computes the errors in tensor elements. Algorithm 5 computes the errors in eigenvalues. Algorithm 6 computes the errors in eigenvectors. Algorithm 7 computes the errors in FA, RA and C_A. Finally, Algorithm 8 computes the errors in the mean diffusivity.

Algorithm 1 Tensor elements

INPUT: b, S_0, S_i, where $i = 1, \cdots, 6$

OUTPUT: tensor D_{xz}, D_{yz}, D_{xy}, D_{xx}, D_{yy}, D_{zz}

1: $A_i \leftarrow (1/b)\ln(S_0/S_i)$, where $i = 1, \cdots, 6$

2: $k \leftarrow 0.25$

3: $D_{xz} \leftarrow k(A_1 - A_2)$; $D_{yz} \leftarrow k(A_3 - A_4)$; $D_{xy} \leftarrow k(A_5 - A_6)$

4: $D_{xx} \leftarrow k(A_1 + A_2 - A_3 - A_4 + A_5 + A_6)$; $D_{yy} \leftarrow k(-A_1 - A_2 + A_3 + A_4 + A_5 + A_6)$; $D_{zz} \leftarrow k(A_1 + A_2 + A_3 + A_4 - A_5 - A_6)$

Algorithm 2 Tensor invariants

INPUT: tensor D_{xz}, D_{yz}, D_{xy}, D_{xx}, D_{yy}, D_{zz}

OUTPUT: tensor invariants I_1, I_2, I_3, v, ϕ

1: $I_1 \leftarrow D_{xx} + D_{yy} + D_{zz}$

2: $I_2 \leftarrow D_{xx}D_{yy} + D_{xx}D_{zz} + D_{yy}D_{zz} - (D_{xy}^2 + D_{yz}^2 + D_{xz}^2)$

3: $I_3 \leftarrow D_{xx}D_{yy}D_{zz} + 2D_{xy}D_{yz}D_{xz} - (D_{zz}D_{xy}^2 + D_{yy}D_{xz}^2 + D_{xx}D_{yz}^2)$

4: $v \leftarrow (2I_1^3 - 9I_1 I_2 + 27I_3)/[2(I_1^2 - 3I_2)^{3/2}]$

5: $\phi \leftarrow arccos(v)/3$

Algorithm 3 Eigenvalues

INPUT: tensor invariants I_1, I_2, I_3, ϕ

OUTPUT: eigenvalues $\lambda_1, \lambda_2, \lambda_3$

1: $h \leftarrow \sqrt{I_1^2 - 3I_2}$

2: $\lambda_1 \leftarrow (I_1 + 2h cos(\phi))/3$

3: $\lambda_2 \leftarrow (I_1 + 2h cos(\phi - 2\pi/3))/3$

4: $\lambda_3 \leftarrow (I_1 + 2h cos(\phi + 2\pi/3))/3$

Algorithm 4 Errors in tensor elements

INPUT: b, S_i, σ_i, where $i = 0, \cdots, 6$

OUTPUT: tensor errors σ_{xz}, σ_{yz}, σ_{xy}, σ_{xx}, σ_{yy}, σ_{zz}

1: $k \leftarrow 0.25/b$; $w_i \leftarrow \sigma_i/S_i$, where $i = 0, \cdots, 6$

2: $\sigma_{xz} \leftarrow k\ln[(1 + w_2)/(1 + w_1)]$; $\sigma_{yz} \leftarrow k\ln[(1 + w_4)/(1 + w_3)]$; $\sigma_{xy} \leftarrow k\ln[(1 + w_6)/(1 + w_5)]$

3: $\sigma_{xx} \leftarrow k\ln[(1 + w_3)(1 + w_4)(1 + w_0)(1 + w_0)/(1 + w_1)/(1 + w_2)/(1 + w_5)/(1 + w_6)]$; $\sigma_{yy} \leftarrow k\ln[(1 + w_1)(1 + w_2)(1 + w_0)(1 + w_0)/(1 + w_3)/(1 + w_4)/(1 + w_5)/(1 + w_6)]$; $\sigma_{zz} \leftarrow k\ln[(1 + w_5)(1 + w_6)(1 + w_0)(1 + w_0)/(1 + w_1)/(1 + w_2)/(1 + w_3)/(1 + w_4)]$;

4: $\sigma_{xz} \leftarrow abs(\sigma_{xz})$; $\sigma_{yz} \leftarrow abs(\sigma_{yz})$; $\sigma_{xy} \leftarrow abs(\sigma_{xy})$; $\sigma_{xx} \leftarrow abs(\sigma_{xx})$; $\sigma_{yy} \leftarrow abs(\sigma_{yy})$; $\sigma_{zz} \leftarrow abs(\sigma_{zz})$

Algorithm 5 Errors in eigenvalues

INPUT: tensor $D_{xz}, D_{yz}, D_{xy}, D_{xx}, D_{yy}, D_{zz}$
                  tensor invariants I_1, I_2, I_3, v, ϕ
                  tensor errors $\sigma_{xz}, \sigma_{yz}, \sigma_{xy}, \sigma_{xx}, \sigma_{yy}, \sigma_{zz}$

OUTPUT: eigenvalue errors $\sigma_{\lambda_1}, \sigma_{\lambda_2}, \sigma_{\lambda_3}$

1: $a_1 \leftarrow \sigma_{xx}^2 + \sigma_{yy}^2 + \sigma_{zz}^2$

2: $a_2 \leftarrow (D_{yy}+D_{zz})^2\sigma_{xx}^2+(D_{xx}+D_{zz})^2\sigma_{yy}^2+(D_{xx}+D_{yy})^2\sigma_{zz}^2+4(D_{xy}^2\sigma_{xy}^2+ D_{xz}^2\sigma_{xz}^2 + D_{yz}^2\sigma_{yz}^2)$

3: $a_3 \leftarrow (D_{yy}D_{zz}-D_{yz}^2)^2\sigma_{xx}^2+(D_{xx}D_{zz}-D_{xz}^2)^2\sigma_{yy}^2+(D_{xx}D_{yy}-D_{xy}^2)^2\sigma_{zz}^2+ 4((D_{xz}D_{yz} - D_{zz}D_{xy})^2\sigma_{xy}^2 + (D_{xy}D_{yz} - D_{yy}D_{xz})^2\sigma_{xz}^2 + (D_{xy}D_{xz} - D_{xx}D_{yz})^2\sigma_{yz}^2)$

4: $b_1 \leftarrow 1.5[(2I_1^2-3I_2)(I_1^2-3I_2)^{-3/2} - I_1(2I_1^3-9I_1I_2+27I_3)(I_1^2-3I_2)^{-5/2}]$

5: $b_2 \leftarrow -4.5I_1(I_1^2-3I_2)^{-3/2} + 2.25(2I_1^3-9I_1I_2+27I_3)(I_1^2-3I_2)^{-5/2}$

6: $b_3 \leftarrow 13.5(I_1^2-3I_2)^{-3/2}; k \leftarrow 1/\sqrt{1-v^2}$

7: $c_1 \leftarrow -kb_1; c_2 \leftarrow -kb_2; c_3 \leftarrow -kb_3$

8: $k_1 \leftarrow sin(\phi); k_2 \leftarrow cos(\phi); k_3 \leftarrow sin(\phi-2\pi/3);$

9: $k_4 \leftarrow cos(\phi-2\pi/3); k_5 \leftarrow sin(\phi+2\pi/3); k_6 \leftarrow cos(\phi+2\pi/3)$

10: $h \leftarrow \sqrt{I_1^2-3I_2}$

11: $d_{11} \leftarrow 1/3 + 2I_1k_2/(3h) - 2k_1c_1h/9; d_{12} \leftarrow -k_2/h - 2k_1c_2h/9; d_{13} \leftarrow -2k_1c_3h/9$

12: $d_{21} \leftarrow 1/3 + 2I_1k_4/(3h) - 2k_3c_1h/9; d_{22} \leftarrow -k_4/h - 2k_3c_2h/9; d_{23} \leftarrow -2k_3c_3h/9$

13: $d_{31} \leftarrow 1/3 + 2I_1k_6/(3h) - 2k_5c_1h/9; d_{32} \leftarrow -k_6/h - 2k_5c_2h/9; d_{33} \leftarrow -2k_5c_3h/9$

14: $\sigma_{\lambda_1} \leftarrow (d_{11}^2a_1 + d_{12}^2a_2 + d_{13}^2a_3)^{1/2}$

15: $\sigma_{\lambda_2} \leftarrow (d_{21}^2a_1 + d_{22}^2a_2 + d_{23}^2a_3)^{1/2}$

16: $\sigma_{\lambda_3} \leftarrow (d_{31}^2a_1 + d_{32}^2a_2 + d_{33}^2a_3)^{1/2}$

Algorithm 6 Errors in eigenvectors

INPUT: tensor $D_{xz}, D_{yz}, D_{xy}, D_{xx}, D_{yy}, D_{zz}$

                eigenvalues $\lambda_1, \lambda_2, \lambda_3$

                tensor errors $\sigma_{xz}, \sigma_{yz}, \sigma_{xy}, \sigma_{xx}, \sigma_{yy}, \sigma_{zz}$

                eigenvalue errors σ_{λ_i}, where $i = 1, 2, 3$

OUTPUT: eigenvector errors $\sigma_{e_{x_i}}, \sigma_{e_{y_i}}, \sigma_{e_{z_i}}$, where $i = 1, 2, 3$

1: $U_1 \leftarrow D_{xy}D_{yz} - (D_{yy} - \lambda_i)D_{xz}$; $U_2 \leftarrow D_{xz}D_{yz} - (D_{zz} - \lambda_i)D_{xy}$; $U_3 \leftarrow D_{xz}D_{xy} - (D_{xx} - \lambda_i)D_{yz}$

2: $q_1 \leftarrow D_{yz}^2\sigma_{xy}^2 + D_{xy}^2\sigma_{yz}^2 + D_{xz}^2\sigma_{yy}^2 + D_{xz}^2\sigma_{\lambda_i}^2 + (\lambda_i - D_{yy})^2\sigma_{xz}^2$

3: $q_2 \leftarrow D_{yz}^2\sigma_{xz}^2 + D_{xz}^2\sigma_{yz}^2 + D_{xy}^2\sigma_{zz}^2 + D_{xy}^2\sigma_{\lambda_i}^2 + (\lambda_i - D_{zz})^2\sigma_{xy}^2$

4: $q_3 \leftarrow D_{xy}^2\sigma_{xz}^2 + D_{xz}^2\sigma_{xy}^2 + D_{yz}^2\sigma_{xx}^2 + D_{yz}^2\sigma_{\lambda_i}^2 + (\lambda_i - D_{xx})^2\sigma_{yz}^2$

5: $r_1 \leftarrow \sqrt{U_1^2U_2^2 + U_2^2U_3^2 + U_1^2U_3^2}$

6: $r_2 \leftarrow r_1^3$; $r_3 \leftarrow U_2^2 + U_3^2$; $r_4 \leftarrow U_1^2 + U_3^2$; $r_5 \leftarrow U_1^2 + U_2^2$

7: $t_{11} \leftarrow U_2/r_1 - U_1^2U_2r_3/r_2$; $t_{12} \leftarrow U_1/r_1 - U_2^2U_1r_4/r_2$; $t_{13} \leftarrow -U_1U_2U_3r_5/r_2$

8: $t_{21} \leftarrow -U_1U_2U_3r_3/r_2$; $t_{22} \leftarrow U_3/r_1 - U_2^2U_3r_4/r_2$; $t_{23} \leftarrow U_2/r_1 - U_3^2U_2r_5/r_2$

9: $t_{31} \leftarrow U_3/r_1 - U_1^2U_3r_3/r_2$; $t_{32} \leftarrow -U_1U_2U_3r_4/r_2$; $t_{33} \leftarrow U_1/r_1 - U_3^2U_1r_5/r_2$

10: $\sigma_{e_{x_i}} \leftarrow (t_{11}^2q_1 + t_{12}^2q_2 + t_{13}^2q_3)^{1/2}$

11: $\sigma_{e_{y_i}} \leftarrow (t_{21}^2q_1 + t_{22}^2q_2 + t_{23}^2q_3)^{1/2}$

12: $\sigma_{e_{z_i}} \leftarrow (t_{31}^2q_1 + t_{32}^2q_2 + t_{33}^2q_3)^{1/2}$

Algorithm 7 Errors in FA, RA, and C_A

INPUT: eigenvalues λ_1, λ_2, λ_3
                eigenvalue errors σ_{λ_1}, σ_{λ_2}, σ_{λ_3}

OUTPUT: σ_{FA}, σ_{RA}, σ_{C_A}

1: $L_1 \leftarrow \sqrt{\lambda_1^2 + \lambda_2^2 + \lambda_3^2}$; $L_2 \leftarrow L_1^3$

2: $L_3 \leftarrow \sqrt{(\lambda_1 - \lambda_2)^2 + (\lambda_2 - \lambda_3)^2 + (\lambda_3 - \lambda_1)^2}$; $L_4 \leftarrow \lambda_1 + \lambda_2 + \lambda_3$

3: $L_5 \leftarrow L_4^2$; $L_6 \leftarrow 2\lambda_1 - \lambda_2 - \lambda_3$; $L_7 \leftarrow 2\lambda_2 - \lambda_3 - \lambda_1$; $L_8 \leftarrow 2\lambda_3 - \lambda_2 - \lambda_1$

4: $L_9 \leftarrow L_1 L_3$; $L_{10} \leftarrow L_3/L_2$; $L_{11} \leftarrow L_4 L_3$; $L_{12} \leftarrow L_3/L_5$

5: $F_1 \leftarrow (L_6/L_9 - \lambda_1 L_{10})/\sqrt{2}$; $F_2 \leftarrow (L_7/L_9 - \lambda_2 L_{10})/\sqrt{2}$; $F_3 \leftarrow (L_8/L_9 - \lambda_3 L_{10})/\sqrt{2}$

6: $\sigma_{FA} \leftarrow (F_1^2 \sigma_{\lambda_1}^2 + F_2^2 \sigma_{\lambda_2}^2 + F_3^2 \sigma_{\lambda_3}^2)^{1/2}$

7: $R_1 \leftarrow (L_6/L_{11} - L_{12})/\sqrt{2}$; $R_2 \leftarrow (L_7/L_{11} - L_{12})/\sqrt{2}$; $R_3 \leftarrow (L_8/L_{11} - L_{12})/\sqrt{2}$

8: $\sigma_{RA} \leftarrow (R_1^2 \sigma_{\lambda_1}^2 + R_2^2 \sigma_{\lambda_2}^2 + R_3^2 \sigma_{\lambda_3}^2)^{1/2}$

9: $C_1 \leftarrow 3\lambda_3/L_5$; $C_2 \leftarrow -3(\lambda_1 + \lambda_2)/L_5$

10: $\sigma_{C_A} \leftarrow (C_2^2 \sigma_{\lambda_1}^2 + C_1^2 \sigma_{\lambda_2}^2 + C_2^2 \sigma_{\lambda_3}^2)^{1/2}$

Algorithm 8 Errors in mean difusivity

INPUT: σ_{xx}, σ_{yy}, σ_{zz}

OUTPUT: $\sigma_{D_{av}}$

1: $\sigma_{D_{av}} \leftarrow \sqrt{\sigma_{xx}^2 + \sigma_{yy}^2 + \sigma_{zz}^2}/3$

These algorithms, as we can see, are all of constant in terms of computational time and space for each voxel. They are simple and scalable. The simplicity of the algorithms allows much efficient applications in practice which is superior to any alternative approaches based on various fitting techniques, such as least-squares methods. The scalability offers direct error estimations for any stage of the error propagation chain.

4   Simulation

The previous section provides the algorithms based on a theoretical framework for establishment of analytical expressions. These theoretical results and algorithms can be very useful in studies of the errors in eigenvalues and eigenvectors, in the mean diffusivity and in the anisotropy indices under different SNR conditions and in comparison of sensitivity to noise among FA, RA and C_A.

We studied, for example, a prolate tensor in the corpus callosum of a volunteer dataset to investigate the typical errors in tensor data under different SNR conditions. The signal intensities (S_0 - S_6) were measured to be (in 8 bit resolution) $200, 138, 95, 14, 79, 42$, and 23, respectively, at spatial resolution $2.4 \times 2.4 \times 7\ mm^3$ at 1.5T for $b = 1000\ s/mm^2$, averaged with four measurements for each signal. These values give the following eigenvalues and mean diffusivity: $\lambda_1 = 0.0017\ mm^2/s$, $\lambda_2 = 0.0003\ mm^2/s$, $\lambda_3 = 0.0001\ mm^2/s$, and $D_{av} = 0.0007\ mm^2/s$. The calculated FA, RA, and C_A from these eigenvalues are 0.8711, 0.7155, and 0.8526, respectively. The SNR was defined in the range of $20 - 120$ for the baseline signal S_0 acquired without applying diffusion-weighting gradients. Based on the range of SNR and the signal intensity S_0, the noise level was thus in the range between 1.7 and 10. The noise corresponding to each SNR was applied to all acquired signal data (S_0–S_6) to quantify the errors in the tensor elements, in the tensor eigenvalues and eigenvectors, in the mean diffusivity and in the diffusion anisotropy indices.

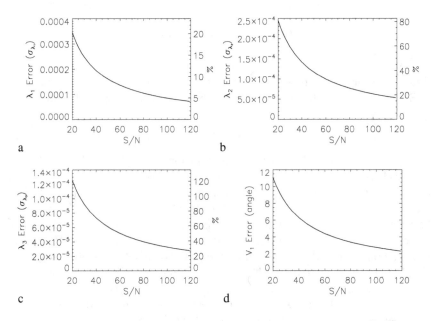

Figure 1. Plots of errors in eigenvalues (a – c), and principal eigenvector (d) as a function of SNR of signal S_0 for a representative voxel in the corpus callosum.

Figure 1 shows the errors in the three eigenvalues (a – c) and in the principal eigenvector (d) in the corpus callosum as a function of SNR of signal S_0. As expected, the errors in the eigenvalues and in the principal eigenvector decrease as SNR increases. This decrease is most rapid for SNR below 70. For SNR at 70, the relative errors on λ_1, λ_2 and λ_3 are 7%, 29% and 44%, respectively. The large relative errors in λ_2 and λ_3 are due to their small values which may occur in such anisotropic diffusion rendering their values imprecise. The error in the orientation of the principal eigenvector at SNR $= 70$ is at an angle of 3.9° away from the expected direction. This error however increases to 11.1° when SNR falls to 20.

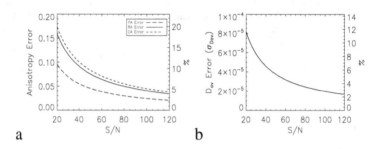

Figure 2. Plots of errors in FA, RA and C_A (a), and in D_{av} (b), as a function of SNR of signal S_0 for a representative voxel in the corpus callosum.

Figure 2(a) shows the errors in the diffusion anisotropy indices (FA, RA and C_A) as a function of SNR of signal S_0. The relative error for FA is the smallest among these three anisotropy indices for any given SNR. The relative error for C_A is larger than that of RA at any given SNR. The gap between FA and RA at any given SNR in the plots is much larger than that between RA and C_A, reflecting a significant error reduction from RA to FA but a small error reduction from C_A to RA. For SNR at 70, the errors on FA, RA and C_A relative to the true FA value are 3.9%, 6.5% and 7.2%, respectively. Figure 2(b) shows the errors in the mean diffusivity D_{av} as a function of SNR of signal S_0. The errors in the mean diffusivity D_{av} decrease as SNR increases as expected. For SNR at 70, the relative error on D_{av} is about 4%.

Figure 3 shows the error in D_{av} as a function of error in FA. The error in FA ranges from 0.02 to 0.095 corresponding to the SNR ranging from 120 to 20 for signal S_0. A linear relationship is clearly seen between the error in mean diffusivity D_{av} and error in FA. This is consistant with recent

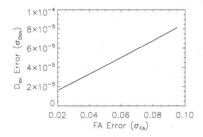

Figure 3. Plot of errors in D_{av} against error in FA.

experimental result in [29].

5   Conclusions

We have established analytical expressions and algorithms for the noise induced errors in the entire tensor computational chain, including the diffusion
tensor elements, the tensor eigenvalues and eigenvectors, the mean diffusivity, and the diffusion anisotropy indices. These mathematical expressions
and algorithms can be used to determine errors for any SNR conditions. In
particular these expressions and algorithms can be used to evaluate errors in
principal eigenvector orientation which has implications for diffusion tensor
tractography.

  These expressions and algorithms offer a simple and efficient way to
demonstrate the noise effects on tensor estimation and provide a means to
quantify the errors for any tensor parameters or tensor derived quantities
in minimum and constant computation time. With these analytical expressions and algorithms, one can easily compare different anisotropy indices in
terms of sensitivity to noise. For example, as confirmed in our simulation
study, FA has the highest and C_A the lowest immunity to noise among FA,
RA and C_A.

  The established theoretical analysis can in fact easily be extended to
determine the errors for any tensor derived parameters at any given SNR.
The approach provides a better insight into the magnitude of noise induced
errors as they evolve through the tensor computational chain and can be
interrogated at any stage of the chain.

  Our algorithms present a number of advantages over numerical approaches
based on least-squares and bootstrap methods. First, the direct analytical
approach removes the heuristic requirement to numerical determination in
conventional approaches based on the least-squares methods. Secondly, our
algorithms are more efficient in terms of computational time and space.
The time complexity is reduced to a constant time for each voxel. This is
better than both the least-squares and bootstrap approaches which require
substantial computation time to be spent on assessment of DTI data quality [43]. Thirdly, our analytical expressions and algorithms are useful for
decision making processes in clinical practice.

BIBLIOGRAPHY

[1] Basser PJ, Mattiello J, and Le Bihan D. MR diffusion tensor spectroscopy and imaging. *Biophys J*, 66:259–267, 1994.

[2] Basser PJ, Mattiello J, and Le Bihan D. Estimation of the effective self-diffusion tensor from the NMR spin echo. *J Magn Reson B*, 103:247–254, 1994.

[3] Basser PJ and Pierpaoli C. Microstructural and physiological features of tissues elucidated by quantitative-diffusion-tensor MRI. *J Magn Reson B*, 111:209–219, 1996.

[4] Pierpaoli C, Jezzard P, Basser PJ, Barnett A, and Di Chiro G. Diffusion tensor MR imaging of the human brain. *Radiology*, 201:637–648, 1996.

[5] van Gelderen P, de Vleeschouwer MHM, DesPres D, Pekar J, van Zijl PCM, and Moonen CTW. Water diffusion and acute stroke. *Magn Reson Med*, 31:154–163, 1994.

[6] Werring DJ, Clark CA, Barker GJ, Thompson AJ, and Miller DH. Diffusion tensor imaging of lesions and normal-appearing white matter in multiple sclerosis. *Neurology*, 52:1626–1632, 1999.

[7] Jones DK, Lythgoe D, Horsfied MA, Simmons A, Williams SC, and Markus HS. Characterization of white matter damage in ischemic leukoaraiosis with diffusion tensor MRI. *Stroke*, 30:393–397, 1999.

[8] Lim KO, Hedehus M, Moseley M, de Crespigny A, Sullivan EV, and Pfefferbaum A. Compromised white matter tract integrity in schizophrenia inferred from diffusion tensor imaging. *Arch Gen Psychiatry*, 56:367–374, 1999.

[9] Werring DJ, Toosy AT, Clark CA, Parker GJ, Barker GJ, Miller DH, and Thompson AJ. Diffusion tensor imaging can detect and quantify corticospinal tract degeneration after stroke. *J Neurol Neurosurg Psychiatry*, 69:269–272, 2000.

[10] Wieshmann UC, Clark CA, Symms MR, Franconi F, Barker GJ, and Shorvon SD. Reduced anisotropy of water diffusion in structural cerebral abnormalities demonstrated with diffusion tensor imaging. *Magn Reson Imaging*, 17:1269–1274, 1999.

[11] Bozzali M, Falini A, Franceschi M, Cercignani M, Zuffi M, Scotti G, Comi G, and Filippi M. White matter damage in Alzheimer's disease assessed in vivo using diffusion tensor magnetic resonance imaging. *J Neurol Neurosurg Psychiatry*, 72:742–746, 2002.

[12] Gauvain KM, McKinstry RC, Mukherjee P, Perry A, Neil JJ, Kaufman BA, and Hayashi RJ. Evaluating pediatric brain tumor cellularity with diffusion-tensor imaging. *Am J Roentgenol*, 177:449–454, 2001.

[13] Sotak CH. The role of diffusion tensor imaging in the evaluation of ischemic brain. *NMR Biomed*, 15:561–569, 2002.

[14] Horsfield MA and Jones DK. Applications of diffusion-weighted and diffusion tensor MRI to white matter diseases - a review. *NMR Biomed*, 15:570–577, 2002.

[15] Werring DJ, Clark CA, Barker GJ, Miller DH, Parker GJM, Brammer MJ, Bullmore ET, Giampietro VP, and Thompson AJ. The structural and functional mechanisms of motor recovery: complementary use of diffusion tensor and functional magnetic resonance imaging in a traumatic injury of the internal capsule. *J Neurol Neurosurg Psychiatry*, 65:863–869, 1998.

[16] Huppi PS, Maier SE, Peled S, Zientara GP, Barnes PD, Jolesz FA, and Volpe JJ. Microstructural development of human newborn cerebral white matter assessed in vivo by diffusion tensor magnetic resonance imaging. *Pediatr Res*, 44:584–590, 1998.

[17] Pfefferbaum A, Sullivan EV, Hedehus M, Lim KO, Adalsteinsson E, and Moseley M. Age-related decline in brain white matter anisotropy measured with spatially corrected echo-planar imaging tensor imaging. *Magn Reson Med*, 44:259–268, 2000.

[18] Hasan KM, Halphen C, Boska MD, and Narayana PA. Diffusion tensor metrics, T2 relaxation and volumetry of the naturally ageing human caudate nuclei in healthy young and middle-aged adults: possible implications for the neurobiology of human brain ageing and disease. *Magn Reson Med*, 59(1):7–13, 2008.

[19] Mamata H, Jolesz FA, and Maier SE. Characterization of central nervous system structures by magnetic resonance diffusion anisotropy. *Neurochem Int.*, 45(4):553–560, 2004.

[20] Basser PJ and Pierpaoli C. A simplified method to measure the diffusion tensor from seven MR images. *Magn Reson Med*, 39:928–934, 1998.

[21] Ulug AM and van Zijl PC. Orientation-independent diffusion imaging without tensor diagonalization: anisotropy definitions based on physical attributes of the diffusion ellipsoid. *J Magn Reson Imaging*, 9:804–813, 1999.

[22] Hasan KM and Narayana PA. Computation of the mean diffusivity and fractional anisotropy maps without tensor decording and diagonalization: theoretical analysis and experimental validation. *Magn Reson Med*, 50:589–598, 2003.

[23] Akkerman E. Efficient measurement and calculation of MR diffusion anisotropy images using the platonic variance method. *Magn Reson Med*, 49:599–604, 2003.

[24] Pierpaoli C and Basser PJ. Toward a quantitative assessment of diffusion anisotropy. *Magn Reson Med*, 36:893–906, 1996.

[25] Xing D, Papadakis NG, Huang CLH, Lee VM, Carpenter TA, and Hall LD. Optimized diffusion weighting for measurement of apparent diffusion coefficient (ADC) in human brain. *Magn Reson Imaging*, 15:771–84, 1997.

[26] Bastin ME, Armitage PA, and Marshall I. A theoretical study of the effect of experimental noise on the measurement of anisotropy in diffusion imaging. *Magn Reson Imaging*, 16:773–785, 1998.

[27] Skare S, Li T, Nordell B, and Ingvar M. Noise considerations in the determination of diffusion tensor anisotropy. *Magn Reson Imaging*, 18:659–669, 2000.

[28] Hirsch JG, Bock M, Essig M, and Schad LR. Comparison of diffusion anisotropy measurements in combination with the flair-technique. *Magn Reson Imaging*, 117:705–716, 1999.

[29] Hasan KM. A framework for quality control and parameter optimization in diffusion tensor imaging: theoretical analysis and validation. *Magnetic Resonance Imaging*, 25(8):1196–1202, 2007.

[30] Tournier JD, Calamante F, King MD, Gadian DG, and Connelly A. Limitations and requirements of diffusion tensor fiber tracking: an assessment using simulations. *Magn Reson Med*, 47:701–708, 2002.

[31] Jones DK. Determining and visualizing uncertainty in estimates of fiber orientation from diffusion tensor MRI. *Magn Reson Med*, 49:7–12, 2003.

[32] Huang H, Zhang J, van Zijl PC, and Mori S. Analysis of noise effects on DTI-based tractography using the brute-force and multi-ROI approach. *Magn Reson Med*, 52:559–565, 2004.

[33] Poonawalla AH and Zhou XJ. Analytical error propagation in diffusion anisotropy calculations. *J Magn Reson Imaging*, 19:489–498, 2004.

[34] Anderson AW. Theoretical analysis of the effects of noise on diffusion tensor imaging. *Magn Reson Med*, 46:1174–1188, 2001.

[35] Chang LC, Koay CG, Pierpaoli C, and Basser PJ. Variance of estimated DTI-derived parameters via first-order perturbation methods. *Magn Reson Med*, 57:141–149, 2007.

[36] Hasan KM, Alexander AL, and Narayana PA. Does fractional anisotropy have better noise immunity characteristics than relative anisotropy in diffusion tensor MRI? An analytical approach. *Magn Reson Med*, 51:413–417, 2004.

[37] Koay CG, Chang LC, Pierpaoli C, and Basser PJ. Error propagation framework for diffusion tensor imaging via diffusion tensor representations. *IEEE Trans Med Imaging*, 26:1017–1034, 2007.

[38] Bevington PR and Robinson DK. *Data reduction and error analysis for the physical sciences*. New York: McGraw-Hill, Inc., 1992.

[39] Shen Y, Pu I, and Clark C. Analytical expressions for noise propagation in diffusion tensor imaging. In *Proc. Intl. Soc. Mag. Reson. Med. 14*, page 1065, Seatle, USA, May 2006.

[40] Hasan KM, Basser PJ, Parker DL, and Alexander AL. Analytical computation of the eigenvalues and eigenvectors in DT-MRI. *J Magn Reson*, 152:41–47, 2001.

[41] Hartmann S. Computational aspects of the symmetric eigenvalue problem of second order tensors. *Technische Mechanik*, 23:283–294, 2003.

[42] Westin CF, Maier SE, Mamata H, Nabavi A, Jolesz FA, and Kikinis R. Processing and visualization for diffusion tensor MRI. *Med Image Anal*, 6:93–108, 2002.

[43] Heim S, Hahn K, Samann PG, Fahrmeir L, and Auer DP. Assessing DTI data quality using bootstrap analysis. *Magn Reson Med*, 52:582–589, 2004.

Ida M. Pu
Department of Computing
Goldsmiths, University of London
London SE14 6NW, UK
Email: i.pu@gold.ac.uk

Yuji Shen
School of Medicine
University of Birmingham
Birmingham, UK
Email: y.shen.1@bham.ac.uk

Empirical Evaluation of Construction Heuristics for the Multidimensional Assignment Problem

Daniel Karapetyan, Gregory Gutin, and Boris Goldengorin

ABSTRACT. The multidimensional assignment problem (MAP) (abbreviated s-AP in the case of s dimensions) is an extension of the well-known assignment problem. The most studied case of MAP is 3-AP, though the problems with larger values of s have also a number of applications. In this paper we consider four fast construction heuristics for MAP. One of the heuristics is new. A modification of the heuristics is proposed to optimize the access to slow computer memory. The results of computational experiments for several instance families are provided and discussed.

1 Introduction

The Multidimensional Assignment Problem (MAP) (abbreviated s-AP in the case of s dimensions) is a well-known optimization problem with a host of applications (see, e.g., [4, 6, 7] for 'classic' applications and [5, 17] for recent applications in solving systems of polynomial equations and centralized multisensor multitarget tracking). In fact, several applications described in [5, 6, 17] naturally require the use of s-AP for values of s larger than 3.

MAP is an extension of a well-known Assignment Problem (AP) which is exactly the two dimensional case of MAP. While AP can be solved in a polynomial time [14], s-AP for $s > 2$ is NP-hard [8].

For a fixed $s \geq 2$, the s-AP is stated as follows. Let $X_1 = X_2 = \ldots = X_s = \{1, 2, \ldots, n\}$. We will consider only vectors that belong to the Cartesian product $X = X_1 \times X_2 \times \ldots \times X_s$. Each vector $e \in X$ is assigned a non-negative weight $w(e)$. For a vector $e \in X$, the component e_j denotes its jth coordinate, i.e., $e_j \in X_j$. A collection of $t \leq n$ vectors $e^1, e^2, \ldots, e^t$ is a *(feasible) partial assignment* if $e^i_j \neq e^k_j$ holds for each $i \neq k$ and $j \in \{1, 2, \ldots, s\}$. The *weight* of a partial assignment A is $w(A) = \sum_{i=1}^{t} w(e^i)$. An *assignment* (or *full assignment*) is a partial assignment with n vectors. The objective is to find an assignment of minimum weight.

The 3-AP is the most studied case of MAP so far. Aiex et al. introduce a Greedy Randomized Adaptive Search Procedure for 3-AP in [1]; an exact algorithm for 3-AP is proposed by Balas and Saltzman in [4]; Crama and Spieksma discuss some special cases of 3-AP and propose approximation algorithms for them. A memetic approach is tried by Huang and Lim in [11]. The more general case of s-AP for arbitrary values of s is less studied. The most recent research by Gutin, Goldengorin and Huang overviews the previous results and discusses the worst case analysis of several MAP construction heuristics [10].

2   Heuristics

There are three construction heuristics for MAP known from the literature: Greedy, Max-Regret [4, 5], and ROM [10]. In this paper, we propose a modification of ROM, Shift-ROM, and compare all four heuristics with respect to solution quality and running time.

2.1   Greedy heuristic

The Greedy heuristic starts with an empty partial assignment $A = \emptyset$. On each of n iterations Greedy finds a vector $e \in X$ of minimum weight, such that $A \cup \{e\}$ is a feasible partial assignment, and adds it to A.

The time complexity of Greedy heuristic is $O(n^s + (n-1)^s + \ldots + 2^s + 1) = O(n^{s+1})$ (if the Greedy algorithm is implemented via sorting of all the vectors according to their weights, the algorithm complexity is $O(n^s \cdot \log n^s)$ however this implementation is inefficient, see Subsection 3.1).

2.2   Max-Regret

The Max-Regret heuristic was first introduced in [4] for 3-AP and its modifications for s-AP were considered in [5].

Max-Regret proceeds as follows. Initialize partial assignment $A = \emptyset$. Set $V_d = \{1, 2, \ldots, n\}$ for each $1 \le d \le s$. For each dimension d and each coordinate value $v \in V_d$ consider every vector $e \in X'$ such that $e_d = v$, where $X' \subset X$ is the set of 'available' vectors, i.e., $A \cup \{e\}$ is a feasible partial assignment if and only if $e \in X'$. Find two vectors e^1_{min} and e^2_{min} in the considered subset $Y_{d,v} = \{e \in X' : e_d = v\}$ such that $e^1_{min} = \operatorname{argmin}_{e \in Y_{d,v}} w(e)$, and $e^2_{min} = \operatorname{argmin}_{e \in Y_{d,v} \setminus \{e^1_{min}\}} w(e)$. Select the pair (d, v) that corresponds to the maximum difference $w(e^2_{min}) - w(e^1_{min})$ and add the vector e^1_{min} for the selected (d, v) to A.

The time complexity of Max-Regret is $O(s \cdot n^s + s \cdot (n-1)^s + \ldots + s \cdot 2^s + s) = O(s \cdot n^{s+1})$.

2.3 ROM

The *Recursive Opt Matching* (ROM) is introduced in [10] as a heuristic of large domination number (see [10] for definitions and results in domination analysis). ROM proceeds as follows. Initialize the assignment A with the trivial vectors: $A^i = (i, i, \ldots, i)$. On each jth iteration of the heuristic, $j = 1, 2, \ldots, s - 1$, calculate an $n \times n$ matrix $M_{i,v} = \sum_{e \in Y(j,i,v)} w(e)$, where $Y(j, i, v)$ is a set of all vectors $e \in X$ such that the first j coordinates of the vector e are equal to the first j coordinates of the vector A^i and the $(j+1)$th coordinate of e is v: $Y(j, i, v) = \{e \in X : e_k = A^i_k, 1 \leq k \leq j$ and $e_{j+1} = v\}$. Let permutation π be a solution of the 2-AP for the matrix M. Set $A^i_{j+1} = \pi(i)$ for each $1 \leq i \leq n$.

The time complexity of ROM heuristic is $O((n^s + n^3) + (n^{s-1} + n^3) + \ldots + (n^2 + n^3)) = O(n^s + sn^3)$.

2.4 Shift-ROM

A disadvantage of the ROM heuristic is that it is not symmetric with respect to the dimensions. For example, if the vector weights do not depend significantly on the last coordinate then the algorithm is likely to work badly. Shift-ROM is intended to solve this problem by trying ROM for different permutations of the instance dimensions. However, we do not wish to try all $s!$ possible dimension permutations as that would increase the running time of the algorithm quite significantly and instead we use only s different permutations: $(X_1 X_2 \ldots X_s)$, $(X_s X_1 X_2 \ldots X_{s-1})$, $(X_{s-1} X_s X_1 X_2 \ldots X_{s-2})$, $\ldots$, $(X_2 X_3 \ldots X_s X_1)$.

In other words, on each run Shift-ROM applies ROM to the problem; upon completion, it renumbers the dimensions for the next run in the following way: $X_1 := X_2$, $X_2 := X_3$, $\ldots$, $X_{s-1} := X_s$, $X_s := X_1$. After s runs, the best solution is selected.

The time complexity of Shift-ROM heuristic is $O((n^s + sn^3) \cdot s) = O(sn^s + s^2 n^3)$.

2.5 Time complexity comparison

Now we can gather the information about the time complexity of the considered heuristics. The following table shows the time complexity of each of the heuristics for different values of s.

	Greedy	Max-Regret	ROM	Shift-ROM
Arbitrary s	$O(n^{s+1})$	$O(sn^{s+1})$	$O(n^s + sn^3)$	$O(sn^s + s^2 n^3)$
Fixed $s = 3$	$O(n^4)$	$O(n^4)$	$O(n^3)$	$O(n^3)$
Fixed $s \geq 4$	$O(n^{s+1})$	$O(n^{s+1})$	$O(n^s)$	$O(n^s)$

3   Performance notes

Modern computer architecture is complex and, hence, not every operation takes the same time to execute. In a standard computer model it is assumed that all the operations take approximately the same time. We will use a more sophisticated model in our further discussion. The idea is to differentiate fast and low memory access operations.

The weight matrix of a MAP instance is normally stored in the Random Access Memory (RAM) of the computer. RAM's capacity is large enough for the very large instances, e.g., nowadays RAM of a common desktop PC is able to hold the weight matrix for 3-AP with $n = 700$, i.e., $3.43 \cdot 10^8$ weights[1]. RAM is a fast storage; one can load gigabytes of data from RAM in one second. However, RAM has a comparatively high latency, i.e., it takes a lot of time for the processor to access even a small portion of data in RAM. The processor cache is intended to minimize the time spent by the processor for waiting for RAM response.

The processor cache exploits two heuristics: firstly, if some data was recently used then there is a high probability it will be used again soon and, secondly, the data is used successively, i.e., if some portion of data is used now then it is probably that the successive portion of data will be used soon. As an example, consider an in place vector multiplication algorithm: on every iteration the algorithm loads a value from the memory, multiplies it and saves the result at the same memory position. So, the algorithm accesses every portion of data twice and the data is accessed successively, i.e., the algorithm accesses the first element of the data, then it accesses the second element, the third etc.

Processor cache[2] is a temporary data storage, relatively small and fast, usually located on the same chip as the processor. It contains several *cache lines* of the same size; each cache line holds a copy of some fragment of the data stored in RAM. Each time the processor needs to access some data in RAM it checks whether this data is already presented in the cache. If this is the case, it accesses this data in the cache instead. Otherwise a 'miss' is detected, the processor suspends, some cache line is freed and a new portion of data is loaded from RAM to cache. Then the processor resumes and accesses the data in the cache as normally. Note that in case of a 'miss' the system loads the whole cache line that is currently 64 bytes on the most of the modern computers [2] and this size tends to grow with the development of computer architecture. Thus, if a program accesses some value in the memory several times in a short period of time it is very

---

[1] Here and further we assume that every weight is represented with a 4 byte integer. The calculations are provided for 2 Gb of RAM.

[2] We provide a simplified overview of cache, for detailed information, see, e.g., [3].

probably that this data will be loaded from RAM just once and then will be stored in the cache so the access time will be minimal. Moreover, if some value is accessed and, thus, loaded from RAM to the processor cache, it is probably that the next value is also loaded since the cache line is large enough to store several values.

With respect to MAP heuristics, there are two key rules for improving the memory subsystem performance:

1. The successive access to the weight matrix (scan), i.e., access to the matrix in the order of its alignment in the memory, is strongly preferred (we use the row-major order [13] for weight matrix in our implementation of the algorithms). Note that if an algorithm accesses, e.g., every second weight in the matrix and does it in the right order, the real complexity of this scan for the memory subsystem is the same as the complexity of a full scan since loading of one value causes loading of several neighbor values.

2. One should minimize the number of the weight matrix scans as much as possible. Even a partial matrix scan is likely to access much more data than the processor cache is able to store, i.e., the data will be loaded from RAM all over again for every scan.

Following these rules may significantly improve the running time of the heuristics. In our experiments, the benefit of following these rules was a speedup of roughly speaking 2 to 5 times.

3.1 Greedy heuristic optimization

A common implementation of the greedy heuristic for combinatorial optimization problem involves sorting of all the weights in the problem. In case of MAP this approach is inefficient since we actually need only n vectors from the n^s set. Another natural implementation of the Greedy heuristic is to scan all available vectors and to choose the lightest one on each iteration but it is very unfriendly with respect to the memory subsystem: it performs n scans of the weight matrix.

We propose a combination of these approaches; our algorithm proceeds as follows. Let $A = \emptyset$ be a partial assignment and B an array of vectors. While $|A| < n$, i.e., A is not a full assignment, the following is repeated. We scan the weight matrix to fill array B with k vectors corresponding to k minimal weights in non-decreasing order: if the weight of the current vector is less than the largest weight in B then we insert the current vector to B in an appropriate position and, if necessary, remove the last element of B. Then, for each vector $e \in B$, starting from the lightest, we check whether $A \cup \{e\}$ is a feasible partial assignment and, if so, add e to A.

Note, that during the second and further cycles we scan not the whole weight matrix but only a subset $X' \subset X$ of the vectors that can be included into the partial assignment A with the feasibility preservation: $A \cup \{x\}$ is a partial assignment for any $x \in X'$. The size of the array B is calculated as $k = \min\{64, |X'|\}$. The constant 64 is obtained from experiments.

The algorithm is especially efficient on the first iterations, i.e., in the hardest part of its work, while the most of the vectors are feasible. However, there exists a bad case for this heuristic. Assume that the weight matrix contains a lot of vectors of the minimal weight $w_{\min}$. Then the array B will be filled with vectors of the weight $w_{\min}$ at the beginning of the scan and, thus, it will contain a lot of similar vectors (recall that the weight matrix is stored in the row-major order and only the last coordinates are varied at the beginning of the scan, so all the vectors processed at the beginning of the scan are likely to have the same first coordinates). As a result, selecting the first of these vectors will cause infeasibility for the other vectors in B. We use an additional heuristic to decrease the running time of the Greedy algorithm for such instances. Let $w_{\min}$ be the minimum possible weight: $w_{\min} = \min_{e \in X'} w(e)$ (sometimes this value is known like for Random instance family it is 1, see Section 4, or one can assume that $w_{\min} = -\infty$). If it occurs during the matrix scan that all the vectors in B have the weight $w_{\min}$, i.e., $w(B_i) = w_{\min}$ for every $1 \leq i \leq k$, then the rest of the scan can be skipped because there is certainly no vector lighter than the maximum weight vector in B. Moreover, it is safe to update $w_{\min}$ with the maximum weight of a vector in B every time before the next matrix scan.

3.2   Max-Regret heuristic optimization

The Max-Regret heuristic naturally requires $O(n^2 s)$ weight matrix partial scans. Each of these scans fixes one coordinate and, thus, every available vector $e \in X'$ (see Subsection 3.1) is accessed s times during each iteration, and this access is very inefficient when the last coordinate is fixed (recall that the weight matrix is stored in a row-major order and, thus, if the last coordinate is fixed then the algorithm accesses every nth value in the memory, i.e., the access is very non-successive and one can assume that this scan will load the whole weight matrix from RAM to cache). In our more detailed computer model, the time complexity of the non-optimized Max-Regret is $O((s-1) \cdot n^{s+1} + n^{s+2})$.

We propose another way to implement Max-Regret. Let us scan the whole available vectors set X' on each iteration. Let L be an $n \times s$ matrix of the lightest vector pairs: $L_{i,j}^1$ and $L_{i,j}^2$ are the lightest vectors when the jth coordinate is fixed as i, and $w(L_{i,j}^1) \leq w(L_{i,j}^2)$. To fill the matrix L we do

the following: for every vector $e \in X'$ and for every coordinate $1 \le d \le s$
check: if $w(e) < w(L^1_{e_d,d})$, set $L^2_{e_d,d} = L^1_{e_d,d}$ and $L^1_{e_d,d} = e$. Otherwise if
$w(e) < L^2_{e_d,d}$, set $L^2_{e_d,d} = e$. Thus, we update the $L_{e_d,d}$ item of the matrix
with the current e if $w(e)$ is small enough. Having the matrix L, we can
easily find the coordinate d and the fixed value v such that $w(L^2_{v,d}) - w(L^1_{v,d})$
is maximized. The vector $L^1_{v,d}$ is added to the solution and the next iteration
of the algorithm is executed.

The proposed algorithm performs just n partial scans of the weight matrix. The matrix L is usually small enough to fit in the processor cache,
so the access to L is fast. Thus, the time complexity of the optimized
Max-Regret in our more detailed computer model is $O(n^{s+1})$.

3.3   ROM heuristic optimization

The ROM heuristic can be implemented in a very friendly with respect to the
memory access way. On the first iteration it fixes the first two coordinates
(n^2 combinations) and enumerates all vectors with these fixed coordinates.
Thus, it scans the whole weight matrix successively. On the next iteration it
fixes three coordinates (n^2 combinations as the second coordinate depends
on the first one), and enumerates all vectors with these fixed coordinates.
Thus, it scans n^2 solid n^{s-3}-size fragments of the weight matrix; further
iterations are similar. As a result, the time complexity of ROM in our more
detailed computer model is the same as in a simple one: $O(n^s + sn^3)$.

3.4   Shift-ROM heuristic optimization

The Shift-ROM heuristic is an extension of ROM; it simply runs ROM s
times, starting it from different coordinates. However, not every run of
ROM is efficient as a part of Shift-ROM. Let us consider the case when
the first iteration of ROM fixes the last two coordinates. For each of the
n^2 combinations of the last two coordinate values, the heuristic scans the
whole weight matrix with the step n^2 between the accessed weights, i.e.,
the distance between the successively accessed weights in the memory is n^2
elements, which is very inefficient. A similar situation occurs when the first
and the last dimensions are fixed.

To avoid this disadvantage, we propose the following algorithm. Let M^d
be an $n \times n$ matrix for every $1 \le d \le s$. Initialize $M^d_{i,j} = 0$ for every
$1 \le d \le s$ and $1 \le i, j \le n$. For each vector $e \in X$ and for each $1 \le d \le s$
set $M^d_{e_d,e_{d+1}} = M^d_{e_d,e_{d+1}} + w(e)$ (here we assume that $e_{s+1} = e_1$). Now the
matrices M^d can be used for the first iteration of every ROM run.

When applying this technique, only one full matrix scan is needed for
the heuristic and this scan is successive. There are several other inefficient
iterations like fixing of the last three coordinates but they influence the

performance insignificantly.

4   Test bed

In this paper we consider four instance families.

Random instance family is a family of random instances, i.e., $w(e)$ is chosen arbitrary for each $e \in X$. Each weight is a uniformly distributed integer number in the interval $[1, 100]$. This instance family is used in [1, 4, 16] and some other papers.

Composite instance family is a family of semi-random instances. They were introduced by Crama and Spieksma for 3-AP as the T problem [7]. We extend this family for s-AP case.

In [7] the 3-AP problem is interpreted as follows. Given a complete tripartite graph $K = \big(X_1 \cup X_2 \cup X_3, (X_1 \times X_2) \cup (X_1 \times X_3) \cup (X_2 \times X_3)\big)$, find a subset A of n triangles, $A \subset X_1 \times X_2 \times X_3$, such that every element of $X_1 \cup X_2 \cup X_3$ occurs in exactly one triangle of A, and the total weight of all the edges covered by triangles A is minimized. The weight of a triangle is calculated as the sum of weights of its edges; the weight of an edge $(i, j) \in X_1 \times X_2$ is $d^1_{i,j}$, the weight of an edge $(i, j) \in X_2 \times X_3$ is $d^2_{i,j}$, and the weight of an edge $(i, j) \in X_1 \times X_3$ is $d^3_{i,j}$, where d^1, d^2, and d^3 are random $n \times n$ matrices of non-negative numbers. In our interpretation of the problem, $w(i_1, i_2, i_3) = d^1_{i_1,i_2} + d^2_{i_2,i_3} + d^3_{i_1,i_3}$.

We introduce an extension of the T problem from [7]. Let us consider a graph $G\big(X_1 \cup X_2 \cup \ldots \cup X_s, (X_1 \times X_2) \cup (X_2 \times X_3) \cup \ldots \cup (X_{s-1} \times X_s) \cup (X_1 \times X_s)\big)$, where the weight of an edge $(i, j) \in X_1 \times X_2$ is $d^1_{i,j}$, the weight of an edge $(i, j) \in X_2 \times X_3$ is $d^2_{i,j}$, ..., and the weight of an edge $(i, j) \in X_{s-1} \times X_s$ is $d^{s-1}_{i,j}$, the weight of an edge $(i, j) \in X_1 \times X_s$ is $d^s_{i,j}$ and d^1, d^2, ..., d^s are random $n \times n$ matrices of non-negative numbers distributed uniformly in the interval $[1, 100]$. The objective is to find a set of n vertex-disjoint s-cycles $C \subset X_1 \times X_2 \times \ldots \times X_s$ such that the total weight of all edges covered by the cycles C is minimized. In our interpretation of the problem, $w(e) = d^1_{e_1,e_2} + d^2_{e_2,e_3} + \ldots + d^{s-1}_{e_{s-1},e_s} + d^s_{e_1,e_s}$.

Additional conditions are applied in [7] to the random matrices d^1, d^2, and d^3, but we do not use these restrictions.

CS instance set is the instance set used by Crama and Spieksma in [7] for the $T\Delta$ problem that is a special case of T, i.e., CS is a subset of the Composite instance family. There are three types of instances, 6 instances per each type: 3 instances of size 33 and 3 instances of size 66. All the instances are of 3-AP. The CS instances meet the triangle inequality, i.e., $d^l(i, j) \le$

$d^l(i, k) + d^l(k, j)$ for every $l \in \{1, 2, 3\}$ and every $i, j, k \in \{1, 2, \ldots, n\}$. For detailed information, see [7].

GP instance family contains pseudo-random instances with the predefined solutions. Predefined instances are generated by an algorithm described by Grundel and Pardalos in [9]. The generator is naturally designed for s-AP for arbitrary large values of s and n. The GP generator is relatively slow and, thus, it was impossible to experiment with large GP instances.

All the instances for this paper are generated with the standard Microsoft .NET random generator [15] which is based on the Donald E. Knuth's subtractive random number generator algorithm [12]. For the seed of the random number sequence we use the following number: $seed = s + n + i$, where i is the index of the instance of this type, $i \in \{1, 2, \ldots, 10\}$. The GP generator is implemented in C++ programming language and, thus, the standard Visual C++ random number generator is used instead; the seed for it is calculated in the same way. The generator for GP instances is available on the web (http://www.ici.ro/camo/forum/grudel/map.txt). The CS instances and solutions are taken from http://www.econ.kuleuven.ac.be/public/NDBAE03/instancesEJOR.htm.

5 Experimental results

We have conducted a number of experiments for the optimized versions of the Greedy, Max-Regret, ROM, and Shift-ROM heuristic (see Section 3). The test bed is discussed in Section 4.

Every experiment, except the experiments with CS instances, includes 10 runs for each of the heuristics; so, 10 instances are produced for every experiment. The evaluation platform is based on AMD Athlon 64 X2 3.0 GHz processor.

The headers in the tables below are as follows:

s is the number of dimensions of the instance.

n is the linear size of the instance, i.e., $n = |X_1| = |X_2| = \ldots = |X_s|$.

Best is the average for the best known objective values for the corresponding instances.

Opt. is the best objective value of the instance. This header is applicable to CS and GP instances only.

Solution error, % is the average value, in percent, over the optimal solution: $error = (value - opt)/opt \cdot 100\%$, where $value$ is the objective value obtained by the heuristic and opt is the optimal objective value. For the instance families where the optimal objective value is unknown, the *Best* value (see above) is used instead.

Running time, ms is the average running time, in milliseconds.

Gr is for Greedy.

M-R is for Max-Regret.

R is for ROM.

S-R is for Shift-ROM.

The results of the experiments with the Random instance family are presented in Table 1. One can see that the solution quality of all the construction heuristics is very poor; the error exceeds 200% over the optimum value on average for every heuristic. (Note that the best values reported for the Random instances are equal or very close to the minimum possible objective values, i.e., to n, and, thus, are equal or very close to the optimal objective values; recall that the minimum weight of every vector in Random instance family is 1.)

Shift-ROM outperforms other heuristics with respect to solution quality on average. For some instances Max-Regret performs better but this is at the cost of much larger running times. Greedy is approximately 100 times faster than Shift-ROM and 2000 times faster than Max-Regret (one can assume that the speedup heuristic in the Greedy implementation works well in this case since Random instances have a lot of vectors of the minimum possible weight) but it is not much worse than the other heuristics with respect to solution quality.

The results of experiments with the Composite instance family are presented in Table 2. The solution quality here is much better than in the previous experiments. Max-Regret produces the best solutions for 3-AP while Shift-ROM is the best for s-AP for $s \geq 4$. Both Shift-ROM and especially Max-Regret are slow; the fastest heuristic is ROM and it produces relatively good solutions. Greedy is slower and produces worse solutions than the ROM heuristic for the Composite instances.

Table 3 contains the experimental results for the CS instance set. Note that CS contains instances of three types; the instances of the same type are grouped together in the table. One can see that the quality of the Shift-ROM heuristic is almost always better than the quality of all other heuristics. ROM solution quality is close to Shift-ROM solution quality and it outperforms Max-Regret and Greedy with respect to both solution quality and running time.

For the GP instance family (Table 4) Max-Regret and Shift-ROM show the best solution quality; the average error for both heuristics is about 10%. However, Shift-ROM maximum error never exceeds 16.1% while Max-Regret error reaches up to 25.7%, and Max-Regret is approximately 10 times slower than Shift-ROM. ROM is the fastest heuristic for GP and it produces only 1.5 times worse solutions than Max-Regret and Shift-ROM.

6 Conclusion

The comparison of the construction heuristics considered in this paper shows that the selection of a particular heuristic depends on the instance set and the quality/time requirements. Greedy is a very fast heuristic for the Random instance family; ROM and Shift-ROM perform well for the Composite instances due to their dimensionwise nature. However, in most of the cases Max-Regret and Shift-ROM, in particular, produce the best solutions. Moreover, Shift-ROM is more stable than Max-Regret with respect to both solution quality and running time. The ROM heuristic operates significantly faster than Shift-ROM at the price of relatively small solution quality decrease. The Greedy heuristic is fast but usually produces the worst solutions.

BIBLIOGRAPHY

[1] R. M. Aiex, M. G. C. Resende, P. M. Pardalos, and G. Toraldo. GRASP with path relinking for three-index assignment. *INFORMS J. on Computing*, 17(2):224–247, 2005.

[2] AMD. *Software Optimization Guide for AMD64 Processors*. Advanced Micro Devices, 2005. http://www.amd.com/us-en/assets/content_type/white_papers_and_tech_docs/25112.pdf.

[3] A. Bailey. Processor cache 101: How cache works. *AMD Developer Central, Technical Articles*, 2006. http://developer.amd.com/TechnicalArticles/Articles/Pages/1128200684.aspx.

[4] E. Balas and M. J. Saltzman. An algorithm for the three-index assignment problem. *Oper. Res.*, 39(1):150–161, 1991.

[5] H. Bekker, E. P. Braad, and B. Goldengorin. Using bipartite and multidimensional matching to select the roots of a system of polynomial equations. In *Computational Science and Its Applications - ICCSA 2005*, volume 3483 of *Lecture Notes Comp. Sci.*, pages 397–406. Springer, 2005.

[6] R. E. Burkard and E. Çela. Linear assignment problems and extensions. In Z. Du and P. Pardalos, editors, *Handbook of Combinatorial Optimization*, pages 75–149. Dordrecht, 1999.

[7] Y. Crama and F. C. R. Spieksma. Approximation algorithms for three-dimensional assignment problems with triangle inequalities. *European Journal of Operational Research*, 60(3):273–279, 1992.

[8] M. R. Garey and D. S. Johnson. *Computers and Intractability: A Guide to the Theory of NP-Completeness (Series of Books in the Mathematical Sciences)*. W. H. Freeman, January 1979.

[9] D. A. Grundel and P. M. Pardalos. Test problem generator for the multidimensional assignment problem. *Comput. Optim. Appl.*, 30(2):133–146, 2005.

[10] G. Gutin, B. Goldengorin, and J. Huang. Worst case analysis of max-regret, greedy and other heuristics for multidimensional assignment and traveling salesman problems. *Journal of Heuristics*, 14(2):169–181, 2008.

[11] G. Huang and A. Lim. A hybrid genetic algorithm for the three-index assignment problem. *European Journal of Operational Research*, 127(1):249–257, July 2006.

[12] D. E. Knuth. *Seminumerical Algorithms*, volume 2 of *The Art of Computer Programming*. Addison-Wesley, Reading, Massachusetts, second edition, 1981.

[13] D. E. Knuth. *Fundamental Algorithms*, volume 1 of *The Art of Computer Programming*, chapter 2.2.6. Addison-Wesley, Reading, Massachusetts, third edition, 1997.

[14] H. W. Kuhn. The hungarian method for the assignment problem. *Naval Research Logistic Quarterly*, 2:83–97, 1955.

[15] Microsoft. *MSDN*, chapter Random Class. Microsoft, 2008. http://msdn2.microsoft.com/en-us/library/system.random.aspx.

[16] W. P. Pierskalla. The multidimensional assignment problem. *Operations Research*, 16:422–431, 1968.

[17] A. J. Robertson. A set of greedy randomized adaptive local search procedure (grasp) implementations for the multidimensional assignment problem. *Comput. Optim. Appl.*, 19(2):145–164, 2001.

Table 1. Heuristics comparison for the Random instance family. Every experiment includes 10 runs.

Inst.	Best	Solution error, %				Running times, ms			
		Gr	M-R	R	S-R	Gr	M-R	R	S-R
3r100	100.0	101.0	18.7	67.3	58.2	7	814	9	46
3r150	150.0	54.5	28.8	33.6	30.1	14	4 260	26	147
3r200	200.0	42.0	16.8	15.5	13.8	25	13 070	64	367
3r250	250.0	37.4	21.4	8.4	6.4	36	32 043	128	714
3r300	300.0	27.2	13.4	3.8	3.3	39	66 719	184	1 229
3r350	350.0	25.0	16.8	2.3	1.8	49	113 559	292	1 889
3r400	400.0	22.2	7.9	1.5	0.9	55	192 683	417	2 827
3r450	450.0	20.9	8.7	0.6	0.3	68	309 674	635	4 115
4r20	20.8	208.7	185.6	310.6	261.5	1	29	1	11
4r30	30.0	206.3	193.3	229.7	203.3	3	204	6	48
4r40	40.0	188.8	118.8	199.8	158.5	4	825	18	140
4r50	50.0	105.2	93.0	128.2	118.4	7	2 650	51	349
4r60	60.0	107.7	98.0	115.2	100.0	8	6 355	94	713
4r70	70.0	85.6	68.7	91.4	85.6	9	13 176	151	1 357
4r80	80.0	74.0	48.8	76.4	71.1	12	27 285	278	2 214
4r90	90.0	38.3	60.1	70.2	57.7	12	45 485	390	3 746
5r10	10.5	545.7	459.0	641.9	363.8	1	12	1	8
5r15	15.0	359.3	338.0	405.3	333.3	2	130	6	59
5r20	20.0	232.0	221.5	332.0	235.5	2	663	23	237
5r25	25.0	229.2	156.8	249.2	219.2	3	2 448	69	719
5r30	30.0	228.7	220.7	230.0	185.3	4	7 057	161	1 718
5r35	35.0	155.1	137.1	194.6	165.7	5	16 552	337	3 707
6r6	6.5	838.5	923.1	912.3	452.3	0	5	0	5
6r9	9.0	674.4	548.9	558.9	387.8	1	63	4	49
6r12	12.0	372.5	392.5	440.8	315.8	2	417	22	292
6r15	15.0	336.7	388.0	408.0	313.3	2	1 599	83	1 037
6r18	18.0	321.1	271.1	372.8	282.8	3	5 778	211	3 003
7r4	4.3	1134.9	1337.2	651.2	393.0	0	2	0	2
7r6	6.0	788.3	768.3	851.7	446.7	1	28	2	33
7r8	8.0	726.3	761.3	626.3	403.8	1	218	16	239
7r10	10.0	722.0	663.0	541.0	355.0	1	1 176	78	986
7r12	12.0	420.8	479.2	475.0	345.8	2	4 788	261	3 757
8r4	4.0	927.5	1185.0	1162.5	510.0	0	6	1	9
8r6	6.0	615.0	688.3	790.0	473.3	1	184	14	218
8r8	8.0	548.8	477.5	723.8	433.8	1	1 987	145	2 097
All avg.		329.2	326.1	340.6	222.5	11	24 913	119	1 088
3-AP avg.		41.3	16.6	16.6	14.3	37	91 603	219	1 417
4-AP avg.		126.8	108.3	152.7	132.0	7	12 001	124	1 072
5-AP avg.		291.7	255.5	342.2	250.5	3	4 477	99	1 075
6-AP avg.		508.6	504.7	538.6	350.4	2	1 572	64	877
7-AP avg.		758.5	801.8	629.0	388.9	1	1 242	71	1 003
8-AP avg.		697.1	783.6	892.1	472.4	1	726	53	775

Table 2. Heuristics comparison for the Composite instance family. Every experiment includes 10 runs.

Inst.	Best	Solution error, %				Running times, ms			
		Gr	M-R	R	S-R	Gr	M-R	R	S-R
3c100	1396.8	43.4	27.2	40.0	33.8	15	870	9	60
3c150	1760.2	38.8	22.3	33.4	30.1	62	4 391	27	162
3c200	2017.8	37.2	19.1	35.6	33.5	158	12 979	68	401
3c250	2276.1	30.4	17.0	35.1	33.6	292	31 738	129	749
3c300	2551.4	27.5	13.2	34.7	32.6	557	69 101	211	1 229
3c350	2696.4	30.4	13.1	37.9	36.8	916	125 919	314	2 023
3c400	3008.5	27.0	9.7	34.6	33.2	1 424	211 784	480	2 925
3c450	3222.1	24.6	9.2	35.1	33.4	2 017	338 248	685	4 282
4c20	875.7	40.8	32.7	24.8	16.7	2	34	1	10
4c30	930.1	51.3	38.0	27.0	21.2	10	222	6	48
4c40	1040.0	50.3	41.2	32.3	27.2	30	919	19	145
4c50	1139.1	58.7	40.4	38.1	30.3	83	2 700	47	356
4c60	1251.0	53.0	35.6	32.8	27.0	154	6 760	93	721
4c70	1360.7	48.5	33.9	31.5	27.0	287	14 678	173	1 332
4c80	1449.5	47.8	33.7	30.8	27.5	543	28 037	284	2 259
4c90	1544.9	45.2	25.4	29.3	25.3	825	51 083	457	3 672
5c10	812.6	38.2	26.0	12.1	7.9	2	14	1	8
5c15	923.1	41.7	33.4	19.8	10.6	9	147	7	57
5c20	988.8	53.2	40.2	21.2	16.3	36	697	24	226
5c25	1026.0	54.2	46.9	26.6	19.5	117	2 577	71	699
5c30	1091.5	63.0	47.5	30.2	23.7	276	7 168	168	1 666
5c35	1171.9	60.5	47.8	28.1	20.8	583	18 804	381	3 680
6c6	817.6	23.5	21.0	9.0	3.5	1	5	1	5
6c9	911.4	33.2	28.8	11.1	6.1	7	66	4	52
6c12	1025.9	38.4	31.4	15.1	9.9	39	412	23	273
6c15	1011.1	44.4	39.4	19.7	14.3	136	1 956	99	989
6c18	1073.7	49.5	44.1	18.5	14.8	387	6 657	246	2 919
7c4	757.4	16.5	12.0	5.2	1.2	0	2	0	2
7c6	940.9	29.0	23.4	5.6	2.3	4	29	2	32
7c8	1000.4	29.3	26.7	12.9	5.1	25	248	17	240
7c10	1086.1	36.8	34.2	11.0	5.6	124	1 278	92	998
7c12	1159.3	45.9	39.5	15.2	8.4	438	5 048	289	3 703
8c4	841.3	15.9	12.5	4.0	1.0	1	6	1	9
8c6	1074.1	26.8	24.0	4.5	1.8	23	183	14	218
8c8	1148.3	34.2	31.7	10.6	4.5	203	2 039	129	2 069
All avg.		39.7	29.2	23.2	18.5	280	27 051	131	1 092
3-AP avg.		32.4	16.4	35.8	33.4	680	99 379	240	1 479
4-AP avg.		49.4	35.1	30.8	25.3	242	13 054	135	1 068
5-AP avg.		51.8	40.3	23.0	16.5	171	4 901	109	1 056
6-AP avg.		37.8	32.9	14.7	9.7	114	1 819	75	848
7-AP avg.		31.5	27.2	10.0	4.5	118	1 321	80	995
8-AP avg.		25.6	22.7	6.4	2.4	76	743	48	765

Table 3. Heuristics comparison for the CS instance family.

Inst.	n	Opt.	Solution error, %				Running times, ms			
			Gr	M-R	R	S-R	Gr	M-R	R	S-R
3DA99N1	33	1608.0	24.5	19.9	<u>0.6</u>	<u>0.6</u>	0.9	11.4	<u>0.4</u>	1.8
3DA99N2	33	1401.0	19.3	10.3	1.1	<u>0.8</u>	0.8	11.5	<u>0.4</u>	2.0
3DA99N3	33	1604.0	15.3	15.3	<u>0.3</u>	<u>0.3</u>	0.7	11.0	<u>0.4</u>	1.9
3DA198N1	66	2662.0	23.7	17.0	1.1	<u>0.2</u>	4.9	156.0	<u>2.8</u>	14.5
3DA198N2	66	2449.0	33.1	36.0	2.0	<u>0.9</u>	5.7	187.2	<u>2.8</u>	13.8
3DA198N3	66	2758.0	17.4	26.0	1.6	<u>0.6</u>	4.9	156.0	<u>2.7</u>	17.6
3DIJ99N1	33	4797.0	6.6	4.8	1.8	<u>1.4</u>	2.4	11.3	<u>0.4</u>	2.2
3DIJ99N2	33	5067.0	5.6	3.3	1.9	<u>1.3</u>	2.2	11.2	<u>0.5</u>	2.4
3DIJ99N3	33	4287.0	7.0	6.2	<u>1.3</u>	<u>1.3</u>	1.8	11.3	<u>0.4</u>	2.1
3DI198N1	66	9684.0	6.1	4.4	1.4	<u>0.9</u>	15.7	124.8	<u>2.8</u>	15.6
3DI198N2	66	8944.0	6.9	4.9	2.1	<u>2.1</u>	17.7	140.4	<u>3.0</u>	17.2
3DI198N3	66	9745.0	7.0	6.2	1.8	<u>1.2</u>	16.7	171.6	<u>3.0</u>	15.3
3D1299N1	33	133.0	6.8	5.3	4.5	<u>1.5</u>	0.6	10.5	<u>0.3</u>	1.7
3D1299N2	33	131.0	8.4	3.8	6.1	<u>3.1</u>	0.6	10.3	<u>0.4</u>	1.8
3D1299N3	33	131.0	7.6	<u>3.1</u>	<u>3.1</u>	<u>3.1</u>	0.5	10.3	<u>0.3</u>	1.7
3D1198N1	66	286.0	5.9	<u>2.8</u>	3.1	3.1	4.6	156.0	<u>2.5</u>	13.2
3D1198N2	66	286.0	3.1	3.5	3.1	<u>2.4</u>	3.9	156.0	<u>2.6</u>	13.5
3D1198N3	66	282.0	7.4	<u>2.5</u>	4.3	3.9	4.3	140.4	<u>2.3</u>	13.1
All avg.			11.8	9.7	2.3	<u>1.6</u>	4.9	82.6	<u>1.6</u>	8.4

Table 4. Heuristics comparison for the GP instance family. Every experiment includes 10 runs.

Inst.	Opt.	Solution error, %				Running times, ms			
		Gr	M-R	R	S-R	Gr	M-R	R	S-R
3gp20	98.8	17.2	18.1	18.0	<u>16.1</u>	0.4	1.7	<u>0.1</u>	0.4
3gp30	150.9	<u>11.0</u>	11.9	14.1	13.8	1.0	7.7	<u>0.3</u>	1.4
3gp40	197.4	12.4	<u>10.5</u>	13.9	13.6	2.0	23.4	<u>0.7</u>	3.1
3gp50	251.5	9.9	<u>9.6</u>	12.2	11.3	4.0	59.0	<u>1.6</u>	6.3
3gp60	286.1	8.9	<u>8.4</u>	13.1	12.1	6.4	126.7	<u>1.9</u>	11.9
3gp70	343.0	8.3	<u>7.7</u>	12.1	11.4	10.6	215.3	<u>3.1</u>	16.9
3gp80	403.7	7.6	<u>6.7</u>	11.6	10.7	17.5	368.2	<u>4.7</u>	25.3
3gp90	434.5	7.5	<u>5.9</u>	11.4	10.5	25.9	578.8	<u>6.4</u>	35.8
3gp100	504.4	<u>5.6</u>	5.8	9.8	9.4	40.0	803.4	<u>8.7</u>	47.4
4gp10	51.5	22.5	19.8	17.7	<u>12.6</u>	0.5	1.4	<u>0.1</u>	0.8
4gp15	69.6	16.8	13.9	9.2	<u>7.0</u>	1.6	7.9	<u>0.5</u>	3.3
4gp20	106.1	11.7	10.9	5.3	<u>4.5</u>	5.5	31.5	<u>1.4</u>	10.0
4gp25	132.9	8.1	8.3	3.5	<u>3.1</u>	15.4	94.5	<u>3.2</u>	27.6
4gp30	145.2	8.8	8.9	2.2	<u>1.5</u>	34.5	205.9	<u>6.7</u>	50.6
5gp4	20.1	30.3	23.4	31.8	<u>15.9</u>	0.2	0.3	<u>0.0</u>	0.3
5gp6	26.9	32.0	<u>13.8</u>	21.2	14.5	0.4	0.8	<u>0.2</u>	0.8
5gp8	36.3	28.7	23.1	17.6	<u>12.7</u>	0.9	3.7	<u>0.3</u>	2.8
5gp10	49.6	17.7	10.1	9.5	<u>7.1</u>	2.7	13.2	<u>0.8</u>	8.5
5gp12	66.2	12.8	8.9	9.2	<u>6.3</u>	6.4	36.1	<u>2.0</u>	19.1
6gp4	20.4	23.5	<u>0.0</u>	28.4	14.2	0.3	0.4	<u>0.1</u>	0.5
6gp6	30.3	30.4	<u>8.9</u>	18.8	11.9	1.5	5.4	<u>0.6</u>	4.5
6gp8	41.8	24.6	<u>1.4</u>	13.6	8.6	5.2	33.0	<u>2.3</u>	25.8
7gp2	10.9	47.7	25.7	13.8	<u>5.5</u>	0.1	0.1	<u>0.0</u>	0.2
7gp3	13.8	33.3	<u>10.1</u>	34.8	13.8	0.2	0.3	<u>0.0</u>	0.5
7gp4	20.2	31.7	<u>0.0</u>	30.2	13.4	0.4	1.6	<u>0.2</u>	2.2
7gp5	25.6	27.0	<u>5.9</u>	19.5	8.6	1.4	8.5	<u>0.8</u>	8.7
8gp2	9.9	36.4	10.1	19.2	<u>7.1</u>	0.1	0.1	<u>0.0</u>	0.3
8gp3	16.3	54.0	<u>0.0</u>	23.9	11.7	0.3	0.7	<u>0.2</u>	1.3
8gp4	19.2	21.4	<u>6.8</u>	28.1	8.3	1.0	8.2	<u>0.8</u>	8.8
All avg.		21.0	<u>10.2</u>	16.3	10.2	6.4	91.0	<u>1.6</u>	11.2
3-AP avg.		9.8	<u>9.4</u>	12.9	12.1	12.0	242.7	<u>3.0</u>	16.5
4-AP avg.		13.6	12.4	7.6	<u>5.8</u>	11.5	68.2	<u>2.4</u>	18.5
5-AP avg.		24.3	15.9	17.9	<u>11.3</u>	2.1	10.8	<u>0.7</u>	6.3
6-AP avg.		26.2	<u>3.4</u>	20.3	11.6	2.3	13.0	<u>1.0</u>	10.3
7-AP avg.		34.9	10.4	24.6	<u>10.3</u>	0.5	2.6	<u>0.3</u>	2.9
8-AP avg.		37.2	<u>5.6</u>	23.7	9.0	0.5	3.0	<u>0.3</u>	3.5

Daniel Karapetyan
Department of Computer Science
Royal Holloway University of London
Egham, Surrey TW20 0EX, UK
Email: Daniel.Karapetyan@gmail.com

Gregory Gutin
Department of Computer Science
Royal Holloway University of London
Egham, Surrey TW20 0EX, UK
Email: G.Gutin@rhul.ac.uk

Boris Goldengorin
Department of Econometrics and Operations Research
University of Groningen
P.O. Box 800, 9700 AV Groningen, The Netherlands
Email: B.Goldengorin@rug.nl

Linear-Space Computation of the Edit-Distance between a String and a Finite Automaton

Cyril Allauzen and Mehryar Mohri

ABSTRACT. The problem of computing the edit-distance between a string and a finite automaton arises in a variety of applications in computational biology, text processing, and speech recognition. This paper presents linear-space algorithms for computing the edit-distance between a string and an arbitrary weighted automaton over the tropical semiring, or an unambiguous weighted automaton over an arbitrary semiring. It also gives an efficient linear-space algorithm for finding an optimal alignment of a string and such a weighted automaton.

1 Introduction

The problem of computing the edit-distance between a string and a finite automaton arises in a variety of applications in computational biology, text processing, and speech recognition [8, 10, 18, 21, 14]. This may be to compute the edit-distance between a protein sequence and a family of protein sequences compactly represented by a finite automaton [8, 10, 21], or to compute the error rate of a word lattice output by a speech recognition with respect to a reference transcription [14]. A word lattice is a weighted automaton, thus this further motivates the need for computing the edit-distance between a string and a weighted automaton. In all these cases, an optimal alignment is also typically sought. In computational biology, this may be to infer the function and various properties of the original protein sequence from the one it is best aligned with. In speech recognition, this determines the best transcription hypothesis contained in the lattice.

This paper presents linear-space algorithms for computing the edit-distance between a string and an arbitrary weighted automaton over the tropical semiring, or an unambiguous weighted automaton over an arbitrary semiring. It also gives an efficient linear-space algorithm for finding an optimal alignment of a string and such a weighted automaton. Our linear-space algorithms are obtained by using the same generic shortest-distance algorithm but by carefully defining different queue disciplines. More precisely, our meta-queue disciplines are derived in the same way from an underling queue discipline defined over states with the same level.

The connection between the edit-distance and the shortest distance in a directed graph was made very early on (see [10, 4, 5, 6] for a survey of string algorithms). This paper revisits some of these algorithms and shows that they are all special instances of the same generic shortest-distance algorithm using different queue disciplines. We also show that the linear-space algorithms all correspond to using the same meta-queue discipline using different underlying queues. Our approach thus provides a better understanding of these classical algorithms and makes it possible to easily generalize them, in particular to weighted automata.

The first algorithm to compute the edit-distance between a string x and a finite automaton A as well as their alignment was due to Wagner [25] (see also [26]). Its time complexity was in $O(|x||A|_Q^2)$ and its space complexity in $O(|A|_Q^2|\Sigma| + |x||A|_Q)$, where Σ denotes the alphabet and $|A|_Q$ the number of states of A. Sankoff and Kruskal [23] pointed out that the time and space complexity $O(|x||A|)$ can be achieved when the automaton A is acyclic. Myers and Miller [17] significantly improved on previous results. They showed that when A is acyclic or when it is a *Thompson automaton*, that is an automaton obtained from a regular expression using Thompson's construction [24], the edit-distance between x and A can be computed in $O(|x||A|)$ time and $O(|x| + |A|)$ space. They also showed, using a technique due to Hirschberg [11], that the optimal alignment between x and A can be obtained in $O(|x| + |A|)$ space, and in $O(|x||A|)$ time if A is acyclic, and in $O(|x||A| \log |x|)$ time when A is a Thompson automaton.

The remainder of the paper is organized as follows. Section 2 introduces the definition of semirings, and weighted automata and transducers. In Section 3, we give a formal definition of the edit-distance between a string and a finite automaton, or a weighted automaton. Section 4 presents our linear-space algorithms, including the proof of their space and time complexity and a discussion of an improvement of the time complexity for automata with some favorable graph structure property.

2   Preliminaries

This section gives the standard definition and specifies the notation used for weighted transducers and automata which we use in our computation of the edit-distance.

Finite-state transducers are finite automata [20] in which each transition is augmented with an output label in addition to the familiar input label

[2, 9]. Output labels are concatenated along a path to form an output sequence and similarly input labels define an input sequence. *Weighted transducers* are finite-state transducers in which each transition carries some weight in addition to the input and output labels [22, 12]. Similarly, *weighted automata* are finite automata in which each transition carries some weight in addition to the input label. A path from an initial state to a final state is called an *accepting path*. A weighted transducer or weighted automaton is said to be *unambiguous* if it admits no two accepting paths with the same input sequence.

The weights are elements of a semiring $(\mathbb{K}, \oplus, \otimes, \bar{0}, \bar{1})$, that is a ring that may lack negation [12]. Some familiar semirings are the tropical semiring $(\mathbb{R}_+ \cup \{\infty\}, \min, +, \infty, 0)$ and the probability semiring $(\mathbb{R}_+ \cup \{\infty\}, +, \times, 0, 1)$, where $\mathbb{R}_+$ denotes the set of non-negative real numbers. In the following, we will only consider weighted automata and transducers over the tropical semiring. However, all the results of section 4.2 hold for an unambiguous weighted automaton A over an arbitrary semiring.

The following gives a formal definition of weighted transducers.

DEFINITION 1. A *weighted finite-state transducer* T over the tropical semiring $(\mathbb{R}_+ \cup \{\infty\}, \min, +, \infty, 0)$ is an 8-tuple $T = (\Sigma, \Delta, Q, I, F, E, \lambda, \rho)$ where Σ is the finite input alphabet of the transducer, Δ its finite output alphabet, Q is a finite set of states, $I \subseteq Q$ the set of initial states, $F \subseteq Q$ the set of final states, $E \subseteq Q \times (\Sigma \cup \{\epsilon\}) \times (\Delta \cup \{\epsilon\}) \times (\mathbb{R}_+ \cup \{\infty\}) \times Q$ a finite set of transitions, $\lambda : I \to \mathbb{R}_+ \cup \{\infty\}$ the initial weight function, and $\rho : F \to \mathbb{R}_+ \cup \{\infty\}$ the final weight function mapping F to $\mathbb{R}_+ \cup \{\infty\}$.

We define the *size* of T as $|T| = |T|_Q + |T|_E$ where $|T|_Q = |Q|$ is the number of states and $|T|_E = |E|$ the number of transitions of T.

The weight of a path π in T is obtained by summing the weights of its constituent transitions and is denoted by $w[\pi]$. The weight of a pair of input and output strings (x, y) is obtained by taking the minimum of the weights of the paths labeled with (x, y) from an initial state to a final state.

For a path π, we denote by $p[\pi]$ its origin state and by $n[\pi]$ its destination state. We also denote by $P(I, x, y, F)$ the set of paths from the initial states I to the final states F labeled with input string x and output string y. The weight $T(x, y)$ associated by T to a pair of strings (x, y) is defined by:

$$T(x, y) = \min_{\pi \in P(I, x, y, F)} \lambda(p[\pi]) + w[\pi] + \rho(n[\pi]). \tag{1}$$

Figure 1(a) shows an example of weighted transducer over the tropical semiring.

Weighted automata can be defined as weighted transducers A with identical input and output labels, for any transition. Thus, only pairs of the form

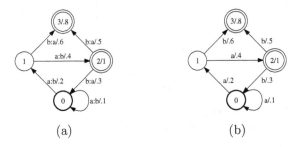

Figure 1. (a) Example of a weighted transducer T. (b) Example of a weighted automaton A. $T(aab, bba) = A(aab) = \min(.1 + .2 + .6 + .8, .2 + .4 + .5 + .8)$. A bold circle indicates an initial state and a double-circle a final state. The final weight $\rho(q)$ of a final state q is indicated after the slash symbol representing q.

(x, x) can have a non-zero weight by A, which is why the weight associated by A to (x, x) is abusively denoted by $A(x)$ and identified with the *weight associated by A to x*. Similarly, in the graph representation of weighted automata, the output (or input) label is omitted. Figure 1(b) shows an example.

3   Edit-distance

We first give the definition of the edit-distance between a string and a finite automaton.

Let Σ be a finite alphabet, and let Ω be defined by $\Omega = (\Sigma \cup \{\epsilon\}) \times (\Sigma \cup \{\epsilon\}) - \{(\epsilon, \epsilon)\}$. An element of Ω can be seen as a symbol edit operation: (a, ϵ) is a deletion, (ϵ, a) an insertion, and (a, b) with $a \neq b$ a substitution. We will denote by h the natural morphism between Ω^* and $\Sigma^* \times \Sigma^*$ defined by $h((a_1, b_1) \cdots (a_n, b_n)) = (a_1 \cdots a_n, b_1 \cdots b_n)$. An *alignment* ω between two strings x and y is an element of Ω^* such that $h(\omega) = (x, y)$.

Let $c : \Omega \to \mathbb{R}_+$ be a function associating a non-negative cost to each edit operation. The cost of an alignment $\omega = \omega_1 \cdots \omega_n$ is defined as $c(\omega) = \sum_{i=1}^{n} c(\omega_i)$.

DEFINITION 2. The *edit-distance* $d(x, y)$ of two strings x and y is the minimal cost of a sequence of symbols insertions, deletions or substitutions transforming one string into the other:

$$d(x, y) = \min_{h(\omega)=(x,y)} c(\omega). \tag{2}$$

When c is the function defined by $c(a, a) = 0$ and $c(a, \epsilon) = c(\epsilon, a) = c(a, b) =$

1 for all a, b in Σ such that $a \neq b$, the edit-distance is also known as the Levenshtein distance. The *edit-distance $d(x, A)$ between a string x and a finite automaton A* can then be defined as

$$d(x, A) = \min_{y \in L(A)} d(x, y), \qquad (3)$$

where $L(A)$ denotes the regular language accepted by A. The *edit-distance $d(x, A)$ between a string x and a weighted automaton A over the tropical semiring* is defined as:

$$d(x, A) = \min_{y \in \Sigma^*} \big(A(y) + d(x, y)\big). \qquad (4)$$

4   Algorithms

In this section, we present linear-space algorithms both for computing the edit-distance $d(x, A)$ between an arbitrary string x and an automaton A, and an optimal alignment between x and A, that is an alignment ω such that $c(\omega) = d(x, A)$.

We first briefly describe two general algorithms that we will use as sub-routines.

4.1   General algorithms

Composition.

The *composition* of two weighted transducers T_1 and T_2 over the tropical semiring with matching input and output alphabets Σ, is a weighted transducer denoted by $T_1 \circ T_2$ defined by:

$$(T_1 \circ T_2)(x, y) = \min_{z \in \Sigma^*} T_1(x, z) + T_2(z, y). \qquad (5)$$

$T_1 \circ T_2$ can be computed from T_1 and T_2 using the composition algorithm for weighted transducers [19, 15]. States in the composition $T_1 \circ T_2$ are identified with pairs of a state of T_1 and a state of T_2. In the absence of transitions with ϵ inputs or outputs, the transitions of $T_1 \circ T_2$ are obtained as a result of the following matching operation applied to the transitions of T_1 and T_2:

$$(q_1, a, b, w_1, q_1') \text{ and } (q_2, b, c, w_2, q_2') \to ((q_1, q_2), a, c, w_1 + w_2, (q_1', q_2')). \quad (6)$$

A state (q_1, q_2) of $T_1 \circ T_2$ is initial (resp. final) iff q_1 and q_2 are initial (resp. final) and, when it is final, its initial (resp.final) weight is the sum of the initial (resp. final) weights of q_1 and q_2. In the worst case, all transitions of T_1 leaving a state q_1 match all those of T_2 leaving state q_2, thus the space and time complexity of composition is quadratic, that is $O(|T_1||T_2|)$.

Shortest distance.

Let A be a weighted automaton over the tropical semiring. The *shortest distance* from p to q is defined as

$$d[p, q] = \min_{\pi \in P(p,q)} w[\pi]. \tag{7}$$

It can be computed using the generic single-source shortest-distance algorithm of [13], a generalization of the classical shortest-distance algorithms. This generic shortest-distance algorithm works with an arbitrary *queue discipline*, that is the order according to which elements are extracted from a queue. We shall make use of this key property in our algorithms. The pseudocode of a simplified version of the generic algorithm for the tropical semiring is given in Figure 2.

SHORTEST-DISTANCE(A, s)

```
1    for  each p ∈ Q do
2          d[p] ← ∞
3    d[s] ← 0
4    S ← {s}
5    while  S ≠ ∅ do
6              q ← HEAD(S)
7              DEQUEUE(S)
8              for  each e ∈ E[q] do
9                    if  (d[s] + w[e] < d[n[e]]) then
10                        d[n[e]] ← d[s] + w[e]
11                        if  (n[e] ∉ S) then
12                              ENQUEUE(S, n[e])
```

Figure 2. Pseudocode of the generic shortest-distance algorithm.

The complexity of the algorithm depends on the queue discipline selected for S. Its general expression is

$$O(|Q| + \mathsf{C}(\mathsf{A}) \max_{q \in Q} \mathsf{N}(\mathsf{q})|E| + (\mathsf{C}(\mathsf{I}) + \mathsf{C}(\mathsf{X})) \sum_{q \in Q} \mathsf{N}(\mathsf{q})), \tag{8}$$

where $\mathsf{N}(\mathsf{q})$ denotes the number of times state q is extracted from queue S, $\mathsf{C}(\mathsf{X})$ the cost of extracting a state from S, $\mathsf{C}(\mathsf{I})$ the cost of inserting a state in S, and $\mathsf{C}(\mathsf{A})$ the cost of an assignment.

With a shortest-first queue discipline implemented using a heap, the algorithm coincides with Dijkstra's algorithm [7] and its complexity is $O((|E| +$

$|Q|) \log |Q|)$. For an acyclic automaton and with the topological order queue discipline, the algorithm coincides with the standard linear-time $(O(|Q| + |E|))$ shortest-distance algorithm [3].

4.2   Edit-distance algorithms

The edit cost function c can be naturally represented by a one-state weighted transducer over the tropical semiring $T_c = (\Sigma, \Sigma, \{0\}, \{0\}, \{0\}, E_c, \bar{1}, \bar{1})$, or T in the absence of ambiguity, with each transition corresponding to an edit operation: $E_c = \{(0, a, b, c(a, b), 0) | (a, b) \in \Omega\}$.

LEMMA 3. *Let A be a weighted automaton over the tropical semiring and let X be the finite automaton representing a string x. Then, the edit-distance between x and A is the shortest-distance from the initial state to a final state in the weighted transducer $U = X \circ T \circ A$.*

Proof. Each transition e in T corresponds to an edit operation $(i[e], o[e]) \in \Omega$, and each path π corresponds to an alignment ω between $i[\pi]$ and $o[\pi]$. The cost of that alignment is, by definition of T, $c(\omega) = w[\pi]$. Thus, T defines the function:

$$T(u, v) = \min_{\omega \in \Omega^*} \{c(\omega) : h(\omega) = (u, v)\} = d(u, v), \qquad (9)$$

for any strings u, v in Σ^*. Since A is an automaton and x is the only string accepted by X, it follows from the definition of composition that $U(x, y) = T(x, y) + A(y) = d(x, y) + A(y)$. The shortest-distance from the initial state to a final state in U is then:

$$\min_{\pi \in P_U(I,F)} w[\pi] = \min_{y \in \Sigma^*} \min_{\pi \in P_U(I,x,y,F)} w[\pi] = \min_{y \in \Sigma^*} U(x, y) \qquad (10)$$

$$= \min_{y \in \Sigma^*} (d(x, y) + A(y)) = d(x, A), \qquad (11)$$

that is the edit-distance between x and A.   ∎

Figure 3 shows an example illustrating Lemma 3. Using the lateral strategy of the 3-way composition algorithm of [1] or an *ad hoc* algorithm exploiting the structure of T, $U = X \circ T \circ A$ can be computed in $O(|x||A|)$ time. The shortest-distance algorithm presented in Section 4.1 can then be used to compute the shortest distance from an initial state of U to a final state and thus the edit distance of x and A. Let us point out that different queue disciplines in the computation of that shortest distance lead to different algorithms and complexities. In the next section, we shall give a queue discipline enabling us to achieve a linear-space complexity.

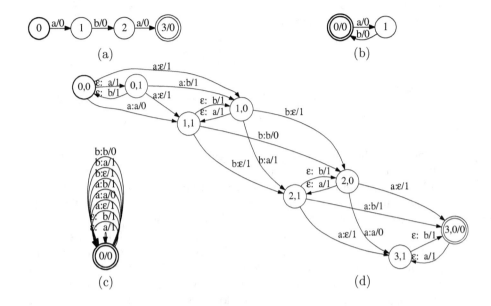

Figure 3. (a) Finite automaton X representing the string $x = aba$. (b) Finite automaton A. (c) Edit transducer T over the alphabet $\{a, b\}$ where the cost of any insertion, deletion and substitution is 1. (d) Weighted transducer $U = X \circ T \circ A$.

4.3   Edit-distance computation in linear space

Using the shortest-distance algorithm described in Section 4.1 leads to an algorithm with space complexity linear in the size of U, i.e. in $O(|x||A|)$. However, taking advantage of the topology of U, it is possible to design a queue discipline that leads to a linear space complexity $O(|x| + |A|)$.

We assume that the finite automaton X representing the string x is topologically sorted. A state q in the composition $U = X \circ T \circ A$ can be identified with a triplet $(i, 0, j)$ where i is a state of X, 0 the unique state of T, and j a state of A. Since T has a unique state, we further simplify the notation by identifying each state q with a pair (i, j). For a state $q = (i, j)$ of U, we will refer to i by the *level* of q. A key property of the levels is that there is a transition in U from q to q' iff $\text{level}(q') = \text{level}(q)$ or $\text{level}(q') = \text{level}(q) + 1$. Indeed, a transition from (i, j) to (i', j') in U corresponds to taking a transition in X (in that case $i' = i+1$ since X is topologically sorted) or staying at the same state in X and taking an input-ϵ transition in T (in that case $i' = i$).

From any queue discipline $\prec$ on the states of U, we can derive a new queue discipline $\prec_l$ over U defined for all q, q' in U as follows:

$$q \prec_l q' \text{ iff } \big(\text{level}(q) < \text{level}(q')\big) \text{ or } \big(\text{level}(q) = \text{level}(q') \text{ and } q \prec q'\big). \quad (12)$$

PROPOSITION 4. *Let $\prec$ be a queue discipline that requires at most $O(|V|)$ space to maintain a queue over any set of states V. Then, the edit-distance between x and A can be computed in linear space, $O(|x| + |A|)$, using the queue discipline $\prec_l$.*

Proof. The benefit of the queue discipline $\prec_l$ is that when computing the shortest distance to $q = (i, j)$ in U, only the shortest distances to the states in U of level i and $i-1$ need to be stored in memory. The shortest distances to the states of level strictly less than $i-1$ can be safely discarded. Thus, the space required to store the shortest distances is in $O(|A|_Q)$.

Similarly, there is no need to store in memory the full transducer U. Instead, we can keep in memory the last two levels active in the shortest-distance algorithm. This is possible because the computation of the out-going transitions of a state with level i only requires knowledge about the states with level i and $i+1$. Therefore, the space used to store the active part of U is in $O(|A|_E + |A|_Q) = O(|A|)$. Thus, it follows that the space required to compute the edit-distance of x and A is linear, that is in $O(|x| + |A|)$. ∎

The time complexity of the algorithm depends on the underlying queue discipline $\prec$. A natural choice is for $\prec$ is the shortest-first queue discipline, that is the queue discipline used in Dijkstra's algorithm. This yields the following corollary.

COROLLARY 5. *The edit-distance between a string x and an automaton A can be computed in time $O(|x||A| \log |A|_Q)$ and space $O(|x| + |A|)$ using the queue discipline $\prec_l$.*

Proof. A shortest-first queue is maintained for each level and contains at most $|A|_Q$ states. The cost for the global queue of an insertion, $\mathsf{C}(\mathsf{I})$, or an assignment, $\mathsf{C}(\mathsf{A})$, is in $O(\log |A|_Q)$ since it corresponds to inserting in or updating one of the underlying level queues. Since $\mathsf{N}(\mathsf{q}) = 1$, the general expression of the complexity (8) leads to an overall time complexity of $O(|x||A| \log |A|_Q)$ for the shortest-distance algorithm. ∎

When the automaton A is acyclic, the time complexity can be further improved by using for $\prec$ the topological order queue discipline.

COROLLARY 6. *If the automaton A is acyclic, the edit-distance between x and A can be computed in time $O(|x||A|)$ and space $O(|x| + |A|)$ using the queue discipline $\prec_l$ with the topological order queue discipline for $\prec$.*

Proof. Computing the topological order for U would require $O(|U|)$ space. Instead, we use the topological order on A, which can be computed in $O(|A|)$, to define the underlying queue discipline. The order inferred by (12) is then a topological order on U.   ∎

Myers and Miller [17] showed that when A is a Thompson automaton, the time complexity can be reduced to $O(|x||A|)$ even when A is not acyclic. This is possible because of the following observation: in a weighted automaton over the tropical semiring, there exists always a shortest path that is *simple*, that is with no cycle, since cycle weights cannot decrease path weight.

In general, it is not clear how to take advantage of this observation. However, a Thompson automaton has additionally the following structural property: a *loop-connectedness* of one. The *loop-connectedness* of A is k if in any depth-first search of A, a simple path goes through at most k back edges. [17] showed that this property, combined with the observation made previously, can be used to improve the time complexity of the algorithm. The results of [17] can be generalized as follows.

COROLLARY 7. *If the loop-connectedness of A is k, then the edit-distance between x and A can be computed in $O(|x||A|k)$ time and $O(|x|+|A|)$ space.*

Proof. We first use a depth-first search of A, identify back edges, and mark them as such. We then compute the topological order for A, ignoring these back edges. Our underlying queue discipline $\prec$ is defined such that a state $q = (i, j)$ is ordered first based on the number of times it has been enqueued and secondly based on the order of j in the topological order ignoring back edges. This underlying queue can be implemented in $O(|A|_Q)$ space with constant time costs for the insertion, extraction and updating operations. The order $\prec_l$ derived from $\prec$ is then not topological for a transition e iff e was obtained by matching a back edge in A and $\text{level}(p[e]) = \text{level}(n[e])$. When such a transition e is visited, $n[e]$ is reinserted in the queue.

When state q is dequeued for the lth time, the value of $d[q]$ is the weight of the shortest path from the initial state to q that goes through at most $l-1$ back edges. Thus, the inequality $\mathsf{N}(\mathsf{q}) \leq k+1$ holds for all q and, since the costs for managing the queue, $\mathsf{C}(\mathsf{I})$, $\mathsf{C}(\mathsf{A})$, and $\mathsf{C}(\mathsf{X})$, are constant, the time complexity of the algorithm is in $O(|x||A|k)$.   ∎

4.4   Optimal alignment computation in linear space

The algorithm presented in the previous section can also be used to compute an optimal alignment by storing a back pointer at each state in U. However, this can increase the space complexity up to $O(|x||A|_Q)$. The use of back pointers to compute the best alignment can be avoided by using a technique due to Hirschberg [11], also used by [16, 17].

As pointed out in previous sections, an optimal alignment between x and A corresponds to a shortest path in $U = X \circ T \circ A$. We will say that a state q in U is a *midpoint* of an optimal alignment between x and A if q belongs to a shortest path in U and $\text{level}(q) = \lfloor |x|/2 \rfloor$.

LEMMA 8. *Given a pair (x, A), a midpoint of the optimal alignment between x and A can be computed in $O(|x| + |A|)$ space with a time complexity in $O(|x||A|)$ if A is acyclic and in $O(|x||A| \log |A|_Q)$ otherwise.*

Proof. Let us consider $U = X \circ T \circ A$. For a state q in U let $d[q]$ denote the shortest distance from the initial state to q, and by $d^R[q]$ the shortest distance from q to a final state. For a given state $q = (i, j)$ in U, $d[(i, j)] + d^R[(i, j)]$ is the cost of the shortest path going through (i, j). Thus, for any i, the edit-distance between x and A is $d(x, A) = \min_j (d[(i, j)] + d^R[(i, j)])$.

For a fixed i_0, we can compute both $d[(i_0, j)]$ and $d^R[(i_0, j)]$ for all j in $O(|x||A| \log |A|_Q)$ time (or $O(|x||A|$ time if A is acyclic) and in linear space $O(|x| + |A|)$ using the algorithm from the previous section forward and backward and stopping at level i_0 in each case. Running the algorithm backward (exchanging initial and final states and permuting the origin and destination of every transition) can be seen as computing the edit-distance between x^R and A^R, the *mirror images* of x and A.

Let us now set $i_0 = \lfloor |x|/2 \rfloor$ and $j_0 = \text{argmin}_j(d[(i_0, j)] + d^R[(i_0, j)])$. It then follows that (i_0, j_0) is a midpoint of the optimal alignment. Hence, for a pair (x, A), the running-time complexity of determining the midpoint of the alignment is in $O(|x||A|)$ if A is acyclic and $O(|x||A| \log |A|_Q)$ otherwise. ∎

The algorithm proceeds recursively by first determining the midpoint of the optimal alignment. At step 0 of the recursion, we first find the midpoint (i_0, j_0) between x and A. Let x^1 and x^2 be such that $x = x^1 x^2$ and $|x^1| = i_0$, and let A^1 and A^2 be the automaton obtained from A by respectively changing the final state to j_0 in A^1 and the initial state to j_0 in A^2. We can now recursively find the alignment between x^1 and A^1 and between x^2 and A^2.

THEOREM 9. *An optimal alignment between a string x and an automaton A can be computed in linear space $O(|x| + |A|)$ and in time $O(|x||A|)$ if A is acyclic, $O(|x||A| \log |x| \log |A|_Q)$ otherwise.*

Proof. We can assume without loss of generality that the length of x is a power of 2. At step k of the recursion, we need to compute the midpoints for 2^k string-automaton pairs $(x_k^i, A_k^i)_{1 \le i \le 2^k}$. Thus, the complexity of step k is in $O(\sum_{i=1}^{2^k} |x_k^i||A_k^i| \log |A_k^i|_Q) = O(\frac{|x|}{2^k} \sum_{i=1}^{2^k} |A_k^i| \log |A_k^i|_Q)$ since $|x_k^i| = |x|/2^k$ for all i. When A is acyclic, the log factor can be avoided and the equality $\sum_{i=1}^{2^k} |A_k^i| = O(|A|)$ holds, thus the time complexity of step k is in $O(|x||A|/2^k)$. In the general case, each $|A_k^i|$ can be in the order of $|A|$, thus the complexity of step k is in $O(|x||A| \log |A|_Q)$.

Since there are at most $\log |x|$ steps in the recursion, this leads to an over-all time complexity in $O(|x||A|)$ if A is acyclic and $O(|x||A| \log |A|_Q \log |x|)$ in general. ∎

When the loop-connectedness of A is k, the time complexity can be improved to $O(k|x||A| \log |x|)$ in the general case.

5 Conclusion

We presented general algorithms for computing in linear space both the edit-distance between a string and a finite automaton and their optimal alignment. Our algorithms are conceptually simple and make use of existing generic algorithms. Our results further provide a better understanding of previous algorithms for more restricted automata by relating them to shortest-distance algorithms and general queue disciplines.

BIBLIOGRAPHY

[1] C. Allauzen and M. Mohri. 3-way composition of weighted finite-state transducers. In O. Ibarra and B. Ravikumar, editors, *Proceedings of CIAA 2008*, volume 5148 of *Lecture Notes in Computer Science*, pages 262–273. Springer-Verlag Berlin Heidelberg, 2008.

[2] J. Berstel. *Transductions and Context-Free Languages*. Teubner Studienbucher: Stuttgart, 1979.

[3] T. Cormen, C. Leiserson, and R. Rivest. *Introduction to Algorithms*. The MIT Press: Cambridge, MA, 1992.

[4] M. Crochemore, C. Hancart, and T. Lecroq. *Algorithms on Strings*. Cambridge University Press, 2007.

[5] M. Crochemore and W. Rytter. *Text Algorithms*. Oxford University Press, 1994.

[6] M. Crochemore and W. Rytter. *Jewels of Stringology*. World Scientific, 2002.

[7] E. W. Dijkstra. A note on two problems in connexion with graphs. *Numerische Mathematik*, 1:269–271, 1959.

[8] R. Durbin, S. Eddy, A. Krogh, and G. Mitchison. *Biological Sequence Analysis: Probalistic Models of Proteins and Nucleic Acids*. Cambridge University Press, Cambridge, UK, 1998.

[9] S. Eilenberg. *Automata, Languages and Machines*, volume A–B. Academic Press, 1974–1976.

[10] D. Gusfield. *Algorithms on Strings, Trees and Sequences*. Cambridge University Press, Cambridge, UK, 1997.

[11] D. S. Hirschberg. A linear space algorithm for computing maximal common subsequences. *Communications of the ACM*, 18(6):341–343, June 1975.

[12] W. Kuich and A. Salomaa. *Semirings, Automata, Languages*. Number 5 in EATCS Monographs on Theoretical Computer Science. Springer-Verlag, 1986.

[13] M. Mohri. Semiring frameworks and algorithms for shortest-distance problems. *Journal of Automata, Languages and Combinatorics*, 7(3):321–350, 2002.

[14] M. Mohri. Edit-distance of weighted automata: General definitions and algorithms. *International Journal of Foundations of Computer Science*, 14(6):957–982, 2003.

[15] M. Mohri, F. C. N. Pereira, and M. Riley. Weighted automata in text and speech processing. In *Proceedings of the 12th biennial European Conference on Artificial Intelligence (ECAI-96), Workshop on Extended finite state models of language, Budapest, Hungary*. John Wiley and Sons, Chichester, 1996.

[16] E. W. Myers and W. Miller. Optimal alignments in linear space. *CABIOS*, 4(1):11–17, 1988.

[17] E. W. Myers and W. Miller. Approximate matching of regular expressions. *Bulletin of Mathematical Biology*, 51(1):5–37, 1989.

[18] G. Navarro and M. Raffinot. *Flexible pattern matching*. Cambridge University Press, 2002.

[19] F. Pereira and M. Riley. *Finite State Language Processing*, chapter Speech Recognition by Composition of Weighted Finite Automata. The MIT Press, 1997.

[20] D. Perrin. Finite automata. In J. V. Leuwen, editor, *Handbook of Theoretical Computer Science, Volume B: Formal Models and Semantics*, pages 1–57. Elsevier, Amsterdam, 1990.

[21] P. A. Pevzner. *Computational Molecular Biology: an Algorithmic Approach*. MIT Press, 2000.

[22] A. Salomaa and M. Soittola. *Automata-Theoretic Aspects of Formal Power Series*. Springer-Verlag, 1978.

[23] D. Sankoff and J. B. Kruskal. *Time Wraps, String Edits and Macromolecules: The Theory and Practice of Sequence Comparison*. Addison-Wesley, Reading, MA, 1983.

[24] K. Thompson. Regular expression search algorithm. *Communications of the ACM*, 11(6):365–375, 1968.

[25] R. A. Wagner. Order-n correction for regular languages. *Communications of the ACM*, 17(5):265–268, May 1974.

[26] R. A. Wagner and J. I. Seiferas. Correcting counter-automaton-recognizable languages. *SIAM Journal on Computing*, 7(3):357–375, August 1978.

Cyril Allauzen
Google Research
76 Ninth Avenue, New York, NY 10011, US
Email: allauzen@google.com

Mehryar Mohri
Courant Institute of Mathematical Sciences
251 Mercer Street, New York, NY 10012, US
and
Google Research
76 Ninth Avenue, New York, NY 10011, US
Email: mohri@cs.nyu.edu

Lossless Image Compression by Block Matching on Practical Parallel Architectures[1]

LUIGI CINQUE, SERGIO DE AGOSTINO, AND LUCA LOMBARDI

ABSTRACT. Storer suggested that fast encoders are possible for bi-level image lossless compression by showing a square greedy matching heuristic, which can be implemented by a simple hashing scheme. We were able to partition an image into up to a hundred areas and to apply such heuristic independently to each area with no relevant loss of compression effectiveness. We show experimental results with up to 32 processors of a 256 Intel Xeon 3.06 GHz processors machine in Italy (avogadro.cilea.it) on a test set of large topographic bi-level images. The expected speed-up of the compression and decompression times is obtained, achieving parallel running times about twenty-five times faster than the sequential ones. Such an approach is suitable only for small scale parallel systems. Rectangle matching improves the compression performance, but it is slower since it requires $O(M \log M)$ time for a single match, where M is the size of the match. Work-optimal $O(\log M \log n)$ time implementations of lossless image compression by rectangular block matching are shown on the PRAM EREW, where n is the size of the image and M is the maximum size of the match, which can be implemented on practical architectures such as meshes of trees, pyramids and multigrids with no scalability issues. The work-optimal implementations on pyramids and multigrids are possible under some realistic assumptions. Decompression on these architectures is also possible with the same parallel computational complexity.

1 Introduction

Storer suggested that fast encoders are possible for two-dimensional lossless compression by showing a square greedy matching heuristic for bi-level images, which can be implemented by a simple hashing scheme [12]. Rectangle

---

[1]The work was supported by the Italian Research Project PRIN and preliminary results were presented in [2], [3] and [4].

matching improves the compression performance, but it is slower since it requires $O(M \log M)$ time for a single match, where M is the size of the match [11]. Therefore, the sequential time to compress an image of size n by rectangle matching is $\Omega(n \log M)$. However, rectangle matching is more suitable for polylogarithmic time parallel implementations on massively parallel network architectures such as the PRAM EREW, the mesh of trees, the pyramid or the multigrid [9].

The technique is a two-dimensional extension of LZ1 compression [10]. Simple and practical heuristics exist to implement LZ1 compression by means of hashing techniques [1], [13], [14]. The hashing technique used for the two-dimensional extension is even simpler.

Among the different ways of reading an image, we assume that the square or rectangle matching compression heuristic scans an m x m' image row by row (*raster* scan). A 64K table with one position for each possible 4x4 subarray is the only data structure used. All-zero and all-one squares or rectangles are handled differently. The encoding scheme is to precede each item with a flag field indicating whether there is a monochromatic square (rectangle), a match, or raw data. When there is a match, the 4x4 subarray in the current position is hashed to yield a pointer to a copy. This pointer is used for the current greedy match and then replaced in the hash table by a pointer to the current position. As mentioned above, the procedure for computing the largest rectangle match with left upper corners in positions (i, j) and (k, h) takes $O(M \log M)$ time, where M is the size of the match, while the procedure for square greedy matches takes $O(M)$ time. Obviously, these procedures can be used respectively to compute the largest monochromatic rectangle and square in a given position (i, j) as well. If the 4 x 4 subarray in position (i, j) is monochromatic, then we compute the largest monochromatic square or rectangle in that position. Otherwise, we compute the largest match in the position provided by the hash table and update the table with the current position. If the subarray is not hashed to a pointer, then it is left uncompressed and added to the hash table with its current position. We want to point out that besides the proper matches we call a match every square or rectangle of the parsing of the image produced by the heuristic. We also call pointer the encoding of a match. The positions covered by matches are skipped in the linear scan of the image.

The analysis of the running time of these algorithms involves a so called *waste factor*, defined as the average number of matches covering the same pixel. In [11], it is conjectured that the waste factor is less than 2 on realistic image data. Therefore, the square greedy matching heuristic takes linear time while the rectangle greedy matching heuristic takes $O(n \log M)$ time. On the other hand, the decoding algorithms are both linear.

By dealing with square matches, we were able to partition an image into up to a hundred areas and to apply such heuristic independently to each area with no relevant loss of compression effectiveness. We show experimental results with up to 32 processors of a 256 Intel Xeon 3.06 GHz processors machine in Italy (avogadro.cilea.it) on a test set of large topographic bi-level images. The expected speed-up of the compression and decompression times is obtained, achieving parallel running times about twenty-five times faster than the sequential ones. Such an approach is suitable only for small scale parallel systems.

Rectangle matching is more suitable for implementations on massively parallel architectures. Work-optimal parallel coding algorithms for loss-less image compression by rectangular block matching were shown on the PRAM-EREW [5, 8] and the mesh of trees [7], requiring $O(\log M \log n)$ time and $O(n/\log n)$ processors. The design of a parallel decoder was left as an open problem as well as the implementation on even simpler archi-tectures as pyramids and multigrids. By slightly modifying the encoder, new parallel coding and decoding algorithms were shown in [4] still requir-ing $O(\log M \log n)$ time and $O(n/\log n)$ processors on the PRAM-EREW. On the mesh of trees though the decoder required $O(\log^2 n)$ time. In this paper, we show how to implement $O(\log M \log n)$ time, $O(n/\log n)$ proces-sors coding and decoding algorithms on the PRAM EREW, mesh of trees, pyramidal, and multigrid architectures with no scalability issues. The work-optimal implementations on pyramids and multigrids are possible under some realistic assumptions.

In section 2, we provide the experimental results on the parallel machine for the square greedy match heuristic. In section 3, we describe the PRAM EREW encoder and decoder. In section 4, we explain how the parallel en-coder and decoder are implemented on the mesh of trees. In section 5, we explain how the parallel encoder and decoder are implemented on the pyra-mid with the same parallel complexity under some realistic assumptions. Conclusions and future work are given in section 6 where the implementa-tions on the multigrid, which is the simplest of the network architectures mentioned above, are discussed.

2 Compression by Square Block Matching on an Array Architecture

Parallel models have two types of complexity: the interprocessor commu-nication and the input-output mechanism. While the input/output issue is inherent to any sublinear algorithm and has standard solutions, the commu-nication cost of the computational phase after the distribution of the data among the processors and before the output of the final result is obviously algorithm-dependent. So, we need to limit the interprocessor communication and involve more local computation. The simplest model for this phase is, of course, a simple array of processors with no interconnections and, therefore, no communication cost.

Dealing with square matches, we were able to partition an image into up to a hundred areas and to apply the square block matching heuristic independently to each area with no relevant loss of compression effectiveness on both the CCITT bi-level image test set and the bi-level version of the five 4096 x 4096 pixels images shown in Figures 1 and 2. With rectangles we cannot obtain the same performance since the width and the length are shortened while the corresponding pointers are more space consuming than with squares. So we would rather implement the square block matching heuristic on an array of processors.

We experimented the heuristic with up to 32 processors of a 256 Intel Xeon 3.06 GHz processors machine in Italy (avogadro.cilea.it) on such test set of large topographic bi-level images. As mentioned in the introduction, the encoding scheme for the pointers uses a flag field indicating whether there is a monochromatic match (0 for the white ones and 10 for the black ones), a proper match (110) or raw data (111). In order to implement decompression as well on an array of processors, we want to indicate the end of the encoding of a specific area. Therefore, we changed the encoding scheme by associating the flag field 1110 to the raw match so that we can indicate with 1111 the end of the sequence of pointers corresponding to a given area. We obtained the expected speed-up of the compression and decompression times with no loss of compression effectiveness, achieving parallel running times about twenty-five times faster than the sequential ones.

We show in Figures 5 and 6 the compression and decompression times of the block matching heuristic on the bi-level version of the five images shown in Figures 3 and 4, doubling up the number of processors of the avogadro.cilea.it machine from 1 to 32. This means that when 2^k processors are involved, for $1 \leq k \leq 5$, the image is partitioned into 2^k areas and the compression heuristic is applied in parallel to each area, independently. As far as decompression is concerned, each of the 2^k processors decodes the pointers corresponding to a given area. We executed the compression and decompression on each image several times. The variances of both the compression and decompression times were small and we report the greatest running times, conservatively. As it can be seen from the values on the tables, also the variance over the test set is quite small. The decompression times are faster than the compression ones and in both cases we obtain the expected speed-up, achieving parallel running times about twenty-five

Figure 1. Images 1, 2 and 3 (from left to right).

times faster than the sequential ones. The images have the same parallel decompression times with 32 processors. On the other hand, image 2 has the greatest sequential compression and decompression times. The greatest compression time with 32 processors and the smallest sequential decompression time is given by image 5. The smallest compression time with 32 processors is given by all other images. Image 1 has the smallest sequential compression time.

3   The PRAM EREW Encoder and Decoder

Rectangle matching is more suitable for implementations on massively parallel architectures. To achieve sublinear time we partition an m x m' image I in x x y rectangular areas, where x and y are $\Theta(\log^{1/2} n)$, and n is the size of the image. In parallel for each area, one processor applies the sequential parsing algorithm so that in $O(\log M \log n)$ time each area is parsed into rectangles, some of which are monochromatic. Before encoding we wish to compute larger monochromatic rectangles, which requires communication among the processors.

Figure 2. Images 4 and 5.

Image	1 proc.	2 proc.	4 proc.	8 proc.	16 proc.	32 proc.
1	76	39	19	11	6	3
2	81	40	23	11	5	3
3	78	39	24	12	6	3
4	79	44	24	11	5	3
5	77	38	22	10	5	4
Avg.	78.2	40	22.4	11	5.4	3.2

Figure 3. Compression times (cs.).

3.1 Computing the Monochromatic Rectangles

Differently from [5], we compute larger monochromatic rectangles by merging adjacent monochromatic areas without considering those monochromatic rectangles properly contained in some area. In practice, these areas are very small and such limitation has no relevant effect on the compression ratio.

We denote with $A_{i,j}$ for $1 \leq i \leq \lceil m/x \rceil$ and $1 \leq j \leq \lceil m'/y \rceil$ the areas into which the image is partitioned. In parallel for $1 \leq i \leq \lceil m/x \rceil$, if i is odd, a processor merges areas $A_{2i-1,j}$ and $A_{2i,j}$ provided they are monochromatic and have the same color. The same is done horizontally for $A_{i,2j-1}$ and $A_{i,2j}$. At the k-th step, if areas $A_{(i-1)2^{k-1}+1,j}$, $A_{(i-1)2^{k-1}+2,j}$, $\cdots$ $A_{i2^{k-1},j}$, with i odd, were merged, then they will merge with areas $A_{i2^{k-1}+1,j}$, $A_{i2^{k-1}+2,j}$, $\cdots$ $A_{(i+1)2^{k-1},j}$, if they are monochromatic with the same color. The same is done horizontally for $A_{i,(j-1)2^{k-1}+1}$, $A_{i,(j-1)2^{k-1}+2}$, $\cdots$ $A_{i,j2^{k-1}}$, with j odd, and $A_{i,j2^{k-1}+1}$, $A_{i,j2^{k-1}+2}$, $\cdots$ $A_{i,(j+1)2^{k-1}}$. After O($\log M$) steps, the

Image	1 proc.	2 proc.	4 proc.	8 proc.	16 proc.	32 proc.
1	43	22	11	6	4	2
2	44	22	12	7	3	2
3	43	22	15	7	4	2
4	43	30	12	7	3	2
5	41	32	15	6	3	2
Avg.	42.8	25.6	13	6.6	3.4	2

Figure 4. Decompression times (cs.).

procedure is completed and each step takes $O(\log n)$ time and $O(n/\log n)$ processors since there is one processor for each area of logarithmic size. Therefore, the image parsing phase is realized with $O(\log M \log n)$ time and $O(n/\log n)$ processors on the PRAM EREW.

3.2 The Parallel Encoder

We derive the sequence of pointers from the image parsing computed above. In $O(\log n)$ time with $O(n/\log n)$ processors we can identify every upper left corner of a match (proper, monochromatic, or raw) by assigning a different segment of logarithmic length on a row to each processor. Each processor detects the upper left corners on its segment. Then, by parallel prefix we obtain a sequence of pointers decodable by the decompressor paired with the sequential heuristic. However, the decoding of such sequence seems hard to parallelize. In order to design a parallel decoder, it is more suitable to produce the sequence of pointers by a raster scan of each of the areas into which the image was originally partitioned, where the areas are ordered by a raster scan themselves. Then, again we can easily derive the sequence of pointers in $O(\log n)$ time with $O(n/\log n)$ processors by detecting in each of the areas the upper left corners of a match and producing the sequence of pointers by parallel prefix.

As mentioned in the introduction, the encoding scheme for the pointers uses a flag field indicating whether there is a monochromatic rectangle (0 for the white ones and 10 for the black ones), a proper match (110), or raw data (111). For the feasibility of the parallel decoder, we want to indicate the end of the encoding of the sequence of matches with the upper left corner in a specific logarithmic area. Therefore, we change the encoding scheme by associating the flag field 1110 to the raw match so that we can indicate with 1111 the end of the sequence of pointers corresponding to a given area.

Moreover, since some areas could be entirely covered by a monochromatic match 1111 is followed by the index associated with the next area by the raster scan. The pointer of a monochromatic match has fields for the width and the length while the pointer of a proper match also has fields for the coordinates of the left upper corner of the copy in the window. In order to save bits, the value stored in any of these fields is the binary value of the field plus 1 (so, we employ the zero value). This coding technique is more redundant than others previously designed, but its compression effectiveness is still better than the one of the square greedy matching technique.

3.3   The Parallel Decoder

The parallel decoder has three phases. Observe that at each position of the binary sequence encoding the image, we read a subsequence of bits that is either 1111 (recall that the k bits following 1111 provide the area index, where k is the number of bits used to encode it) or can be interpreted as a flag field of a pointer. Then, in the first phase we reduce the binary sequence to a doubly-linked structure and apply the well-known Euler tour technique in order to identify for each area the corresponding pointers. The reduction works as follows: link each position p of the sequence to the position next to the end of the subsequence starting in position p that either can be interpreted as a pointer or is equal to 1111 followed by k bits. For those suffixes of the sequence which can be interpreted as pointers, their first positions are linked to a special node denoting the end of the coding. For those suffixes of the sequence which cannot be interpreted as pointers, their first positions are not linked to anything. The linked structure is a forest with one tree rooted in the special node denoting the end of the coding and the other trees rooted in the first position of a suffix of the encoding sequence not interpretable as a pointer. The first position of the binary sequence is a leaf of the tree rooted in the special node. The sequence of pointers encoding the image is given by the path from the first position to the root. In order to compute such path we need the children to be doubly-linked to the parent. Then, we need to reserve space for each node to store the links to the children. Each node has at most five children since there are only four different pointer sizes (white, black, raw, or proper match). So, for each position p of the binary sequence we set aside five locations $[p, 1] \cdots [p, 5]$, initially set to zero. When a link is added from position p' to p, depending on whether the subsequence starting in position p' is 1111 or can be interpreted as a pointer to a raw, white, black or proper match, the value p' is overwritten on location $[p, 1]$, $[p, 2]$, $[p, 3]$ $[p, 4]$ or $[p, 5]$, respectively. Then, by means of the well-known Euler technique we can linearize the linked structure and apply list ranking to obtain the path from the first position of the sequence to the root of its tree. It is well-known that all this can be computed in $O(\log n)$ time with $O(n/\log n)$ processors on the PRAM EREW, since the row image size is greater than the size of the sequence. Then, still in $O(\log n)$ time with $O(n/\log n)$ processors we can identify the positions on the path corresponding to 1111.

In the second phase of the parallel decoder a different processor decodes the sequence of pointers corresponding to a different area. As far as the pointers to monochromatic matches are considered, each processor decompresses either a match contained in an area or the portion of the match corresponding to the left upper area. Therefore, after the second phase an area might not be decompressed. Obviously, the second phase requires $O(\log n)$ time and $O(n/\log n)$ processors on the PRAM EREW.

The third phase completes the decoding. In the previous subsection, we denoted with $A_{i,j}$ for $1 \leq i \leq \lceil m/x \rceil$ and $1 \leq j \leq \lceil m'/y \rceil$ the areas into which the image was partitioned by the encoder. At the first step of the third phase, one processor for each area $A_{2i-1,j}$ decodes $A_{2i,j}$, if $A_{2i-1,j}$ is the upper left portion of a monochromatic match and the length field of the corresponding pointer informs that $A_{2i,j}$ is part of the match. The same is done horizontally for $A_{i,2j-1}$ and $A_{i,2j}$ (using the width field of its pointer) if it is known already by the decoder that $A_{i,2j-1}$ is part of a monochromatic match. Similarly at the k-th step, one processor for each of the areas $A_{(i-1)2^{k-1}+1,j}, A_{(i-1)2^{k-1}+2,j}, \cdots A_{i2^{k-1},j}$, with i odd, decodes the areas $A_{i2^{k-1}+1,j}, A_{i2^{k-1}+2,j}, \cdots A_{(i+1)2^{k-1},j}$, respectively. The same is done horizontally for $A_{i,(j-1)2^{k-1}+1}, A_{i,(j-1)2^{k-1}+2}, \cdots A_{i,j2^{k-1}}$, with j odd, and $A_{i,j2^{k-1}+1}, A_{i,j2^{k-1}+2}, \cdots A_{i,(j+1)2^{k-1}}$. After $O(\log M)$ steps the image is entirely decompressed. Each step takes $O(\log n)$ time and $O(n/\log n)$ processors since there is one processor for each area of logarithmic size. Therefore, the decoder is realized with $O(\log M \log n)$ time and $O(n/\log n)$ processors on the PRAM EREW.

4   The Mesh of Trees Implementations

A *mesh of trees* is a network of $3N^2 - 2N$ processors with N being a power of 2, consisting of an N x N grid where a complete binary tree of processors is built on each row and each column as shown in Figure 5. First, we give a detailed description of the mesh of trees compression algorithm. Then, we provide the implementation of the parallel decoder.

Let max be equal to max $\{m, m'\}$. We assume m and m' have the same order of magnitude, as in practice with the height and the width of an image. Let N be the smallest power of 2 greater than $\lceil max/\log^{1/2} n \rceil$, where n is the size of an m x m' image. Then, the number of processors of the mesh of trees is $O(n/\log n)$ and we can store the logarithmic rectangular areas

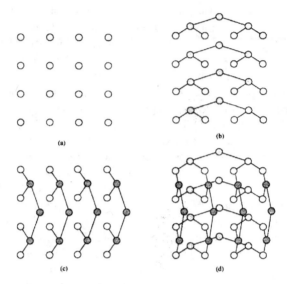

Figure 5. The construction of a mesh of trees with a 4x4 processor grid.

into which the parallel algorithm partitions the image into the N x N grid. Starting from the upper left corner of the grid, a different processor stores a different rectangular area and applies the sequential compression heuristic to such area. The remaining processors are inactive. From now on, we will refer only to the active processors.

4.1 Computing the Monochromatic Rectangles

After the compression heuristic has been executed on each area, it is easy to see that the PRAM EREW procedure to compute larger monochromatic rectangles can be implemented on a mesh of trees with the same number of processors without slowing it down. In fact, if i is odd, the processors storing areas $A_{2i-1,j}$ and $A_{2i,j}$ merge them provided they are monochromatic and have the same color. The same is done horizontally for $A_{i,2j-1}$ and $A_{i,2j}$. At the k-th step, if areas $A_{(i-1)2^{k-1}+1,j}$, $A_{(i-1)2^{k-1}+2,j}$, $\cdots$ $A_{i2^{k-1},j}$, with i odd, were merged, the processor storing area $A_{i2^{k-1},j}$ will broadcast to the processors storing the areas $A_{i2^{k-1}+1,j}$, $A_{i2^{k-1}+2,j}$, $\cdots$ $A_{(i+1)2^{k-1},j}$ to merge with the above areas, if they are monochromatic with the same color. This is done in logarithmic time using the column trees. The same is done horizontally for $A_{i,(j-1)2^{k-1}+1}$, $A_{i,(j-1)2^{k-1}+2}$, $\cdots$ $A_{i,j2^{k-1}}$, with j

odd, and $A_{i,j2^{k-1}+1}, A_{i,j2^{k-1}+2}, \cdots A_{i,(j+1)2^{k-1}}$, using the row trees. After $O(\log M)$ steps, the procedure is completed and each step takes $O(\log n)$ time and $O(n/\log n)$ processors since there is one processor for each area of logarithmic size. Therefore, the image parsing phase is realized with $O(\log M \log n)$ time and $O(n/\log n)$ processors on the mesh of trees.

4.2 The Parallel Encoder

The sequence of pointers for each area can be trivially produced on the grid. With the possibility of a parallel output, the sequences can be put together by parallel prefix on a mesh of trees in $O(\log n)$ time with $O(n/\log n)$ processors.

4.3 The Parallel Decoder

We know that the end of the encoding of an area is indicated by 1111 followed by the index of the area corresponding to the next encoding. Then, we can store the encodings in the positions of the grid corresponding to the locations of the areas in the image. In fact, the first phase of the PRAM EREW decoding algorithm corresponds to the input process of a distributed memory system (as the mesh of trees is) and is not part of our complexity analysis. At this point, each processor on the grid completes the second phase of the decoder described in subsection 2.3. Then, it is easy to see that the third and last phase of the PRAM decoder is implementable on a mesh of trees with the same number of processor and no slowdown. In conclusion, the decoder takes $O(\log M \log n)$ time on a mesh of trees with $O(n/\log n)$ processors.

5 The Pyramid Implementations

An N x N *pyramid* is a network consisting of $\log N + 1$ two-dimensional grids with N being a power of 2, each one of size $N/2^k$ x $N/2^k$ for $0 \leq k \leq \log N$. The grids are interconnected so that the (i, j) processor on the 2^k x 2^k grid is connected to processors $(2i - 1, 2j - 1)$, $(2i - 1, 2j)$, $(2i, 2j - 1)$ and $(2i, 2j)$ on the 2^{k+1} x 2^{k+1} grid, as shown in Figure 6 for the 4 x 4 pyramid network. As for the mesh of trees, let N be the smallest power of 2 greater than $\lceil max/\log^{1/2} n \rceil$, where n is the size of an m x m' image and max is equal to max $\{m, m'\}$. If we assume that m and m' have the same order of magnitude, the number of processors of the pyramid is $O(n/\log n)$ and we can store the logarithmic rectangular areas into which the parallel algorithm partitions the image into the N x N grid. Starting from the upper left corner of the grid, a different processor stores a different rectangular area and applies the sequential compression heuristic to such area while the other processors remain inactive. From now on, we will refer only to the active processors.

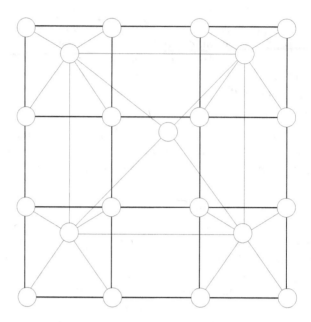

Figure 6. A 4 x 4 pyramid network.

5.1   Computing the Monochromatic Rectangles

After the compression heuristic has been executed on each area, we have
to show how the PRAM EREW procedure to compute larger monochromatic rectangles can be implemented on a pyramid with the same number
of processors without slowing it down. This is possible by making some
realistic assumptions. Let ℓ_R and w_R be the length and the width of a
monochromatic match R, respectively. We define $s_R = \max \{\ell_R, w_R\}$. We
make a first assumption that the number of monochromatic matches R
with $s_R \geq 2^k \lceil \log^{1/2} n \rceil$ is $O(N^2/2^{2k})$ for $1 \leq k \leq \log N - 1$. If i is odd,
the processors storing areas $A_{2i-1,j}$ and $A_{2i,j}$ merge them provided they are
monochromatic and have the same color. The same is done horizontally for
$A_{i,2j-1}$ and $A_{i,2j}$. At the k-th step, if areas $A_{(i-1)2^{k-1}+1,j}$, $A_{(i-1)2^{k-1}+2,j}$,
$\cdots A_{i2^{k-1},j}$, with i odd, were merged for $w_1 \leq j \leq w_2$, the processor storing
area A_{i2^{k-1},w_2} will broadcast to the processors storing the areas $A_{i2^{k-1}+1,j}$,
$A_{i2^{k-1}+2,j}$, $\cdots A_{(i+1)2^{k-1},j}$ to merge with the above areas for $w_1 \leq j \leq w_2$,
if they are monochromatic with the same color. The same is done horizon-
tally, that is, if $A_{i,(j-1)2^{k-1}+1}$, $A_{i,(j-1)2^{k-1}+2}$, $\cdots A_{i,j2^{k-1}}$, with j odd, were
merged for $\ell_1 \leq i \leq \ell_2$, the processor storing area $A_{\ell_2,j2^{k-1}}$ will broadcast

to the processors storing the areas $A_{i,j2^{k-1}+1}$, $A_{i,j2^{k-1}+2}$, $\cdots$ $A_{i,(j+1)2^{k-1}}$ to merge with the above areas for $\ell_1 \leq i \leq \ell_2$, if they are monochromatic with the same color. After $O(\log M)$ steps, the procedure is completed. If the waste factor is less than 2, as conjectured in [11], we can make a second assumption that each pixel is covered by a constant small number of monochromatic matches. It follows from this second assumption that the information about the monochromatic matches is distributed among the processors of a grid in a way very close to uniform. Then, it follows from the first assumption that the amount of information each processor of the grid at level k must broadcast is constant, for $1 \leq k \leq \log N - 1$. Therefore, each step takes $O(\log n)$ time and the image parsing phase is realized with $O(\log M \log n)$ time and $O(n/\log n)$ processors on the pyramid. Finally, we want to point out that the unique processor of the 1 x 1 grid at level $\log n$ is not involved in the computation of the image parsing and is used only for the ineherently sequential input/output operations which have, generally speaking, standard solutions for network algorithms.

5.2 The Parallel Encoder

The sequence of pointers for each area can be trivially produced on the grid at level 0. This is, obviously, realized in $O(\log n)$ time on a pyramid with $O(n/\log n)$ processors.

5.3 The Parallel Decoder

As for the mesh of trees, we can store the encodings of each area in the positions of the grid at level 0 corresponding to the locations of the areas in the image. At this point, each processor on the grid completes the second phase of the decoder described in subsection 2.3. Then, it is easy to see that the third and last phase of the PRAM decoder has the same parallel computational complexity on the pyramid with the same realistic assumptions we made for the coding phase. In conclusion, the decoder takes $O(\log M \log n)$ time on a pyramid with $O(n/\log n)$ processors.

6   Conclusions

Parallel coding and decoding algorithms for lossless image compression by rectangular block matching were shown requiring $O(\log M \log n)$ time and $O(n/\log n)$ processors on the PRAM-EREW, the mesh of trees and the pyramid. The parallel coding algorithms are work-optimal since the sequential time required by the coding is $\Omega(n \log M)$. On the other hand, the parallel decoding algorithms are not work-optimal since the sequential decompression time is linear. The mesh of trees and pyramid implementations of the decoder have the same performance of the PRAM EREW implementation if we do not consider the input process. The pyramid is

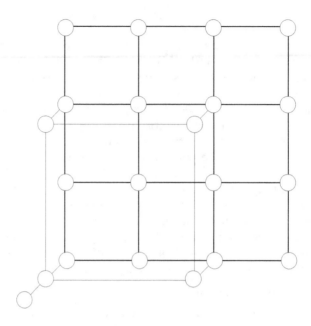

Figure 7. A 4 x 4 multigrid network.

a simpler architecture than the mesh of trees [9] and needs some realistic assumptions to give the same performance. There exist real parallel machines whose architecture is a pyramid. One of them is at the University of Pavia in Italy (PAPIA). As future work, we wish to implement our algorithms on this machine. An even simpler architecture than the pyramid is the multigrid, which is computationally equivalent up to a factor of 2 in speed to the pyramid and it is the simplest among the architectures with small diameter and large bisection width [9]. Since the multigrid is a subgraph of the pyramid (Figure 7), we wish to implement and experiment our algorithms on this architecture as well. On the other hand, compression by square block matching can be implemented locally on a simple array architecture at no communication cost and experimental results with up to 32 processors were obtained, achieving parallel compression and decompression running times about twenty-five times faster than the sequential ones. However, when the number of processors increases the compression ratio of the procedure starts deteriorating. Such loss becomes too relevant with a change in the order of magnitude of the amount of processors. Therefore, this approach is not practical for a massively parallel architecture. The detection of large monochromatic areas is essential to achieve good compression of bi-level images by block matching and communication among processors is necessary on a large scale system. Recently, we have realized that with the assumptions made for the pyramid implementations we can relax on the requirement of an architecture with large bisection width and design a work-optimal implementation of compression by rectangular block matching on a two-dimensional architecture such as a full binary tree [6]. The communication cost might be low enough to realize an efficient implementation on one of the available parallel machines.

BIBLIOGRAPHY

[1] R. P. Brent, "A Linear Algorithm for Data Compression," *Australian Computer Journal* **19** (1987) 64–68.

[2] Cinque L. and De Agostino S. [2008]. "Lossless Image Compression by Block Matching on Practical Massively Parallel Architectures", *Proceedings Prague Stringology Conference*, 26-34.

[3] L. Cinque L., S. De Agostino S. and L. Lombardi [2008]. "Speeding up Lossless Image Compression: Experimental Results on a Parallel Machine", *Proceedings Prague Stringology Conference*, 35-45.

[4] L. Cinque and S. De Agostino, "A Parallel Decoder for Lossless Image Compression by Block Matching," *Proceedings Data Compression Conference* (2007) 183–192.

[5] L. Cinque L., S. De Agostino and F. Liberati, "A Work-Optimal Parallel Implementation of Lossless Image Compression by Block Matching," *Nordic Journal of Computing* **10** (2003) 13-20.

[6] S. De Agostino, "Compressing Bi-Level Images by Block Matching on a Tree Architecture," paper in preparation.

[7] S. De Agostino, "Lossless Image Compression by Block Matching on a Mesh of Trees," *Proceedings Data Compression Conference* (2006) 443.

[8] S. De Agostino, "A Work-Optimal Parallel Implementation of Lossless Image Compression by Block Matching", *Proceedings Prague Stringology Conference* (2002) 2-10.

[9] F. T. Leighton, *Introduction to Parallel Algorithms and Architectures* (Morgan-Kaufmann, 1992).

[10] A. Lempel A. and J. Ziv J., "A Universal Algorithm for Sequential Data Compression," *IEEE Transactions on Information Theory* **23** (1977) 337–343.

[11] J. A. Storer and H. Helfgott, "Lossless Image Compression by Block Matching," *The Computer Journal* **40** (1997) 137-145.

[12] J. A. Storer, "Lossless Image Compression using Generalized LZ1-Type Methods," *Data Compression Conference* (1996) 290-299.

[13] J. R. Waterworth, *Data Compression System,* US Patent 4 701 745.

[14] D. A. Whiting, G. A. George and G. E. Ivey, *Data Compression Apparatus and Method,* US Patent 5016009.

Luigi Cinque
Computer Science Department
Sapienza University
Via Salaria 113, 00198 Rome, Italy
Email: `cinque@di.uniroma1.it`

Sergio De Agostino
Computer Science Department
Sapienza University
Via Salaria 113, 00198 Rome, Italy
Email: deagostino@di.uniroma1.it

Luca Lombardi
Computer Science Department
University of Pavia
Via Ferrara 1, 27100 Pavia, Italy
Email: luca.lombardi@unipv.it

NFA Simulation Using Deterministic State Cache[1]

JAN HOLUB AND TOMÁŠ KADLEC

ABSTRACT. The nondeterministic finite automaton (NFA) cannot
be directly used because of its nondeterminism. There are two ways
to use it: Firstly, transform it to the equivalent deterministic finite
automaton, or secondly, simulate the run of an NFA in a deterministic way. We present a deterministic state cache method that com-
bines these two approaches. It uses the fast memory of already used
deterministic states, which can under certain circumstances dramatically accelerate the basic simulation method, while we can control
the amount of memory used. We present an implementation based
on a hash array.

1 Introduction

Many tasks in Computer Science use or can use finite automata. However, a
non-automaton approach is sometimes preferred in such cases because of the
fear of exponential blow up of the states (memory complexity, preprocessing
time) when transforming a nondeterministic finite automaton (NFA) to the
equivalent deterministic finite automaton (DFA). The transformation uses
the standard subset construction [13, 12] eliminating inaccessible states [7]
(See Algorithm 1.). The resulting DFA can have up to $2^{|Q_{\text{NFA}}|}$ states, where
$|Q_{\text{NFA}}|$ is the number of states of the NFA. This exponential blow up of
states can make the approach unusable. In practice, the increase in states
is not always exponential, but the space complexity and the preprocessing
time are still very considerable.

One can avoid such a blow up by using NFA simulation [7]. There are
several algorithms one can use. The basic simulation method (BSM) [7]
enables us to simulate any general NFA, but with relatively high time and
space complexities. Two specialized methods, called dynamic programming
(DP) [17, 19] and bit parallelism (BP) [6, 18, 3, 20], have been developed

[1]This research has been partially supported by the Ministry of Education, Youth and
Sports under research program MSM 6840770014, and by the Czech Science Foundation
as project No. 201/09/0807.

for tasks in approximate pattern matching. It has been shown [8, 9, 10] not
only that these algorithms simulate the run of an NFA, but also enable these
NFA simulators to be used for other tasks in approximate pattern matching.
These simulation methods have better time and space complexities than
BSM, but due to the cost of these improvements they cannot be applied to
general NFA. They require some kind of regular transition structure. The
selection of a proper simulation method for a given pattern matching task
is discussed in [11].

In BP, the NFA states are organized into levels, each represented by one
bit-vector. The time complexity of BP is given by the number of bitwise
operations each representing some transition. Each of the NFA states for
approximate string matching using Hamming distance, except for rightmost
and bottom line states, has only two outgoing transitions: match (always
leading to the state right of the current state) and replace (always leading
to the state lower right of the current state). Since all match transitions
have the same direction, they can be performed in one bitwise operation for
each level of states, using only one mask vector. The same holds for replace
transitions, but no mask vector is needed. Each new type of transition can
increase the time complexity and the number of mask vectors needed. In
the worst case, we can have τ bit operations and $\tau * |\Sigma|$ mask vectors for an
NFA with τ transitions.

In DP, the NFA states are organized into columns, each represented by
an integer. There cannot be more than one state active in each column, so
DP also cannot be used for any NFA. Moreover loss of the regular structure
of the transitions described in the previous paragraph would lead to a very
complicated NFA simulation code.

Thus the only simulation method applicable to any NFA is BSM. In
this paper we propose a modification of the BSM [7] algorithm utilizing a
Deterministic State Cache (DSC) in order to accelerate BSM. We provide
theoretical basis along with practical implementation. We also try to provide general rules for successful DSC design based on experiments and on a
comparison of DSC with other methods.

2   Definitions

A *nondeterministic finite automaton* is a quintuple $(Q, \Sigma, \delta, q_0, F)$, where Q
is a set of states, Σ is a set of input symbols, δ is a mapping $Q \times (\Sigma \cup \{\varepsilon\}) \mapsto$
$\mathcal{P}(Q)$, $q_0 \in Q$ is an initial state, and $F \subseteq Q$ is a set of final states. We can
extend δ to $\hat{\delta}$ so that it is a mapping $Q \times \Sigma^* \mapsto \mathcal{P}(Q)$ defined as $\hat{\delta}(q, \varepsilon) = \{q\}$;
$\hat{\delta}(q, au) = \hat{\delta}(p, u)$, where $q \in Q$, $p \in \delta(q, a)$, $a \in \Sigma \cup \{\varepsilon\}$, $u \in \Sigma^*$.

$\varepsilon CLOSURE(q) = \{p \mid p \in \hat{\delta}(q, \varepsilon)\}$ is a set of states reachable from state q
using a path consisting of ε-transitions only. Note that $q \in \varepsilon CLOSURE(q)$.

For a set of states $P \subseteq Q$ applies $\varepsilon CLOSURE(P) = \{\bigcup_{q \in P} \varepsilon CLOSURE(q)\}$.

A *deterministic finite automaton* is an NFA with δ being a partial mapping $Q \times \Sigma \hookrightarrow Q$. (Note that DFA is a special case of NFA.)

NFA accepts some input string $w \in \Sigma^*$, if and only if there is a path spelling w leading from the initial state to a final state. Formally it accepts string w if $\hat{\delta}(q_0, w) \cap F \neq \emptyset$. There may be a state with more than one outgoing transition labeled by the same symbol. Thus when reading the input string we have more than one way to choose what the next move will be. That is why we cannot use NFA directly and we have to find other ways to use it in a deterministic way.

One can transform NFA to the equivalent DFA or simulate a run of NFA using BSM (or DP and BP in pattern matching).

3   Determinization vs. simulation

In determinization of NFA (see Algorithm 1.) we consider each subset of NFA states as a DFA state. If q, p are NFA states, q'_1, q'_2 are states of the equivalent DFA such that $p \in q'_1, q \in q'_2$, and a transition for symbol $a \in \Sigma$ leads from q to p, then a transition labeled by a also leads from q'_1 to q'_2.

We start with the $\varepsilon CLOSURE$ of the initial state which forms the initial DFA state q'_0. We consider all transitions leading from NFA states of q'_0 labeled by symbol a. Thus we get another DFA state. We do the same for all symbols of Σ and for all DFA states that we create during the process. Thus we process all NFA transitions, and on the other hand we avoid inaccessible states (states that cannot be reached from the initial state).

The simulation of an NFA run using BSM [7] (see Algorithm 2.) starts in the same way as NFA determinization. However, in this case a set S of active states is computed. Evaluating all transitions labeled by a current input symbol t_i leading from all states in S we compute a new set S of active states.

Actually these sets of active states are just the NFA states composing the DFA states that would result from the determinization of the NFA. The difference is that in NFA simulation we compute only those DFA states that we currently need and we forget information about the current DFA state once we move to the newly computed state to which the current transition leads. Thus we may not compute all transitions and DFA states, but on the other hand we may compute some DFA states more than once. In NFA determinization we compute all transitions and all DFA states and we store them all.

NFA simulation thus has smaller space requirements (DFA states and transitions are not stored) but higher time complexity (the next DFA has to be computed in each step). When DFA is too large to fit in the memory,

---

Algorithm 1 (Transformation of NFA to the equivalent DFA)

Input: NFA $M = (Q, \Sigma, \delta, q_0, F)$
Output: DFA M' accepting language $L(M)$
Method: $M' = (Q', \Sigma, \delta', q_0', F')$, where Q', δ', q_0 are constructed in
                the following way:

1: $q_0' \leftarrow \varepsilon CLOSURE(q_0)$
2: $Q' \leftarrow \emptyset$
3: $S \leftarrow \{q_0'\}$ // not yet processed states
4: **for each** $q' \in S$ **do**
5:    **for each** $a \in \Sigma$ **do**
6:       $\delta'(q', a) \leftarrow \bigcup_{q \in q'} \varepsilon CLOSURE(\delta(q, a))$
7:       **if** $\delta'(q', a) \notin Q'$ and $\delta'(q', a) \notin S$ **then**
8:          $S \leftarrow S \cup \delta'(q', a)$
9:       **end if**
10:   **end for**
11:   **if** $q' \cap F \neq \emptyset$ **then**
12:      $F' \leftarrow F' \cup \{q'\}$
13:   **end if**
14:   $S \leftarrow S \setminus \{q'\}$
15:   $Q' \leftarrow Q' \cup \{q'\}$
16: **end for**

---

we have to use an NFA simulation technique. Also when DFA is complex and the input text is relatively short, it may be faster to use an NFA simulation technique than to construct the whole DFA and then to run it. A resolution system for when to use which technique (especially in pattern matching) is described in [11].

4   Simulation with a deterministic state cache

We combine the two approaches: simulation and determinization. In each step of the automaton run, we check if we already have the transition (defined as a triple: source DFA state, input symbol, and destination DFA state) that we are going to use computed and stored in DSC. If so, we use it and read another input symbol. If such a transition was not found in DSC, we compute the transition using BSM, we store it in DSC, and then we use it. If DSC is full, we use a cache technique to remove a transition in order to make space for the computed transition.

The technique of remembering some computed information is also known as memoization [16], and it is used for example in the LISP programming language. The idea of on-the-fly construction of DFA is also not new. Orig-

---

Algorithm 2 (Simulation of NFA run—Basic Simulation Method)

Input:	NFA $M = (Q, \Sigma, \delta, q_0, F)$, input text $T = t_1 t_2 \ldots t_n$.
Output:	Simulation of run of NFA.
Method:	Set of active states S.

1:   $S \leftarrow \varepsilon CLOSURE(q_0)$
2:   $i \leftarrow 1$
3: **while** $i \leq n$ **and** $S \neq \emptyset$ **do**
4:     $S \leftarrow \bigcup_{q \in S} \varepsilon CLOSURE(\delta(q, t_i))$
5:     **if** $S \cap F \neq \emptyset$ **then**
6:       **print** information connected with reaching final state
7:     **end if**
8:     $i \leftarrow i + 1$
9: **end while**

---

inally it was mentioned in [2], then it was discussed in [4] for approximate string matching using Hamming distance, in [15] for approximate string matching using Levenshtein distance, and for the general problem in [14]. However, we have not found any practical implementation. In addition to the practical implementation of on-the-fly construction of DFA we add the cache of the deterministic states so that we can control the memory used by removing some of the infrequently used deterministic states.

Thus we compute only transitions that are needed, we reuse transitions computed in the past, while we control the amount of memory used by the size of DSC fixed at the beginning of the automaton run. When we use transitions stored in DSC, the algorithm runs almost as fast DFA, otherwise it runs almost as fast as BSM. The optimum case for the algorithm is when a complex automaton most of the time uses a small subset of transitions (that fit in DSC). This usually happens in many useful applications, e.g., pattern matching. The optimum case is obviously when a frequently used part of DSC is stored directly in the CPU cache.

5 Implementation

The implementation of BSM with DSC is based on the implementation of BSM [7] using bit vectors and exploiting bit parallelism.

For an automaton $M = (Q, \Sigma, \delta, q_0, F)$ with ε-transitions and input text $T = t_1 t_2 \ldots t_n$ we construct a transition function δ as a matrix $\mathcal{T}$ of size $|Q| \times |\Sigma|$ of bit vectors each of size $|Q|$ bits. The j-th bit in vector $\mathcal{T}[i, a]$ (denoted as $\mathcal{T}[i, a]_j$) is 1, if $q_j \in \delta(q_i, a)$, or $\mathcal{T}[i, a]_j = 0$, otherwise, where $i \in 0..(|Q|-1)$, $a \in \Sigma$. The set of final states of the automaton is represented by a bit vector $\mathcal{F}$ of size $|Q|$, where $\mathcal{F}_j = 1$, if $q_j \in F$, or $\mathcal{F}_j = 0$, otherwise.

ε-transitions are represented by an array of bit vectors $\mathcal{E}$ of size $|Q|$. In the bit vector $\mathcal{E}[i]$, the j-th bit (denoted as $\mathcal{E}[i]_j$) is 1, if $q_j \in \varepsilon CLOSURE(q_i)$, or $\mathcal{E}[i]_j = 0$, otherwise. The set of active states in step i, $0 \leq i \leq n$ of simulation NFA is represented by a bit vector $\mathcal{S}[i]$ of size $|Q|$, where bit $\mathcal{S}[i]_j = 1$, if state q_j is active in the i-th step of the simulation, or $\mathcal{S}[i]_j = 0$, otherwise.

Algorithm 3 (Simulation of NFA run utilizing DSC–ε-transitions, bit vectors)

Input: NFA, $\mathcal{T}$, $\mathcal{F}$, $\mathcal{E}$, automaton M, input text $T = t_1 t_2 \ldots t_n$.
Output: Simulation of run of NFA.
Method: Set of active states S.

1: $\mathcal{S}[0] \leftarrow \mathcal{E}[0]$ // only $\varepsilon CLOSURE(q_0)$ is active at the beginning
2: $i \leftarrow 1$
3: **while** $i \leq n$ **and** $\mathcal{S}[i-1] \neq [000\ldots0]$ **do**
4:   **if** $\mathcal{S}[i] \leftarrow getFromCache(\mathcal{S}[i-1], t_i)$ is unsuccessful **then**
5:     $\mathcal{S}[i] \leftarrow [000\ldots0]$
6:     **for** $j \leftarrow 0,1,\ldots|Q|-1$ **do**
7:       **if** $\mathcal{S}[i-1]_j = 1$ **then** // q_j is active in $(i-1)$-th step
8:         $\mathcal{S}[i] \leftarrow \mathcal{S}[i]$ OR $\mathcal{T}[j,t_i]$ // transitions from state $q_j * /$
9:       **end if**
10:     **end for**
11:     $\mathcal{S}'[i] \leftarrow [000\ldots0]$
12:     **for** $j \leftarrow 0,1,\ldots|Q|-1$ **do** // computing $\varepsilon CLOSURE(\mathcal{S}[i])$
13:       **if** $\mathcal{S}[i]_j = 1$ **then**
14:         $\mathcal{S}'[i] \leftarrow \mathcal{S}'[i]$ OR $\mathcal{E}[i]$
15:       **end if**
16:     **end for**
17:     $\mathcal{S}[i] \leftarrow \mathcal{S}'[i]$
18:     $putToCache(\mathcal{S}[i-1], t_i, \mathcal{S}[i])$
19:   **end if**
20:   **if** $\mathcal{S}[i]$ AND $\mathcal{F} \neq 0$ **then**
21:     **print** information connected with reaching final state
22:   **end if**
23:   $i \leftarrow i+1$
24: **end while**

When we construct bit vector $\mathcal{S}[i+1]$ holding a new set of active states, we have to evaluate all transitions labeled by symbol t_{i+1} leading from each NFA state q_j which is active in the i-th step of simulation (i.e., $\mathcal{S}[i]_j = 1$). This evaluation can be done at once by exploiting bit parallelism—

we apply bitwise operation OR for bit vectors $\mathcal{S}[i+1]$ and $\mathcal{T}[j, t_{i+1}]$. The implementation of BSM with DSC is shown in Algorithm 3.

Note that lines 5–19 of Algorithm 3 are skipped when the requested transition is found precomputed in the DSC.

Note also that each combination of the bits of vector $\mathcal{S}[i]$ represents a DFA state. If $\mathcal{S}[i] = \mathcal{S}[i']$, then $\mathcal{S}[i]$ and $\mathcal{S}[i']$ represent the same DFA state.

5.1   Cache structure and management policies

Unlike in DFA determinization, the sets of active states in BSM do not have any kind of unique and easily processable identifier. Therefore we have to implement DSC as an associative data structure like that shown in Figure 1. The key in the cache is a source state and a transition label symbol. The value is the destination state and *cmpf*, which is a cache management policy field used by cache management policies to store the value related to the given element.

Key		Value	
bit-vector	alphabet symbol	bit-vector	*cmpf*
110110	s_i	01101	65
100100	s_l	11001	35
...	...	...	...
110110	s_x	01001	99

Figure 1. Deterministic state cache data structure.

The core of DSC are functions *getFromCache*() and *putToCache*(), which differ according to the cache management policy chosen. The *Clock* cache management policy is the naive one. We need to know the cache size, denoted c_s, and to maintain one global cycle counter (variable i can be used for this purpose). When the cache is full we purge the element $(i \bmod c_s)$. A big advantage of the clock method is that the connected overhead for the cache management is insignificant, and it is reduced only to incrementing the counter and once in a while a modulo computation. On the other hand, in the longer run each element has the same chance to be purged and so this policy breaks the principles of temporal and spatial locality—the basic principles ensuring cache efficiency.

The *Least Recently Used* cache management policy (LRU) is a very sophisticated method, because of we select for purging an element that has not been recently used, so the principles of temporal and spatial locality are nicely supported by the LRU method against the clock method. One possible solution is to implement LRU as a combination of two techniques—NFU, which stands for Not Frequently Used, and aging. The NFU technique uses time-stamping—for each element, the frequency of its use is counted. Aging gets around the shortcomings of time-stamping. For each element it lowers its counter of the element slightly in each step.

However, early experiments showed that the overhead connected with aging—the whole cache must be traversed after a cache look-up is performed and the *cmpf* field must be lowered for each record—makes such an approach unusable. As a solution we introduce and implement the LRU-NFU-wA policy, which stands for Least Recently Used—Not Frequently Used—weak Aging, a slightly modified variant of the LRU policy with a dramatically reduced aging overhead. The aging routine is only executed when storing an element into the cache, when the cache must be traversed anyhow to select the element with the smallest *cmpf* to purge. On the other hand, a stored "time-stamp" loses its previous quality and provides a less accurate time statistic. That is why we call this technique weak aging.

The principle of temporal and spatial locality works in various pattern matching automata, since most of the time the automaton uses the initial state and some states in the close neighbourhood. The size of the neighbourhood depends on factors such as the type of the text, size of the alphabet, and the maximum number of errors allowed (in approximate pattern matching).

6   Experiments

We use hash arrays for cache internal data storage. The hash of the key is computed as a hash of the byte-stream into which a bit-vector is serialized and then the current input symbol t_i is appended.

We provide two different implementations of the cache. The first one, denoted BSMCache, is based on generic Google :: dense_hash_map [1], fast hash table implementation, using the open addressing technique with quadratic probing. The second implementation, denoted DSCache, is specialized and is intended only as the DSC for the BSM. Internally it is implemented as a chained hash table. For both implementations the clock and LRU-NFU-wA cache management policies are provided.

Please note that in all presented graphs, the clock or LRU-NFA-wA policy refers to the use of BSMCache with the appropriate policy, while clock-1 or LRU-NFU-wA-1 refers to the use of DSCache with the appropriate policy. Also note that the principles of the cache management policies used here are the same for BSMCache and DSCache, so the hit rate dependencies for the same cache management policies are typically very close or even equal. Therefore the curves that represent these dependencies may overlap, and only the topmost curve is visible.

For benchmarking we chose English text, Alice's Adventures in Wonderland, by Lewis Carroll, and a randomly selected part of the complete genome of the E. Coli bacterium, both of them from the Canterbury Corpus [5]. As patterns, random substrings of the texts were chosen.

We use NFA for approximate string matching using Levenshtein distance up to k. That is, each occurrence found should be converted to the pattern using at most k replace, delete, and insert operations. For more details, see [7]. We compare BSM without DSC, BSM with DSC using various cache management policies described above, and two other simulation methods called bit parallelism (BP) [6, 18, 3, 20] and dynamic programming (DP) [17, 19]. We remark again that BP and DP cannot be used for general NFA but only for NFAs in approximate pattern matching that have a special regular structure of transitions. The experiments presented on one approximate pattern matching task may lead to the conclusion that DP is the best method for NFA in approximate pattern matching. However, the mutual relation of the running times of BSM, DP and BP varies according to the parameters of the pattern matching task: pattern length, maximum number of errors allowed, and size of the alphabet. Selection of the optimum method for a given configuration of the task is discussed in [11].

6.1 Influence of cache size

In Figure 1.2(a) we present the dependency of the run time on the cache size for an NFA for approximate string matching using Levenshtein distance. For cache size greater than 500 both methods offer a huge speed-up, regardless the cache management policy, compared to the original BSM. Note that the clock policy works nicely even without strictly respecting the principles of locality and purging the cache randomly. This is due to its small overhead. However, for a cache being smaller than 500, BSMCache is slower than BSM. The changeover is roughly when the hit rate falls below 0.7, as shown in Figure 1.2(b).

In Figure 1.2(b) we can see that even for a very low hit rate (about 0.2) DSCache brings some speed-up compared to the original BSM, and also to the BSMCache. The overhead connected with the use of DSCache is really small. Naturally, the hit rates remain the same for both implementations, because they differ only in the hash map that is applied, and not in the cache management policies that are provided.

6.2 Dependency on the input length

For this experiment we derived the input files from the original files with lengths ranging from 1 to $\frac{1}{10}$ of the original files. The results are depicted in Figure 3.

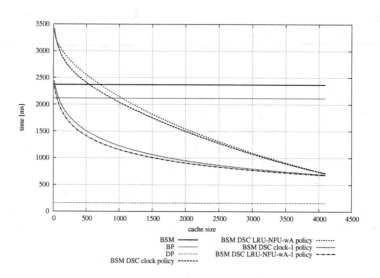

(a) Run time

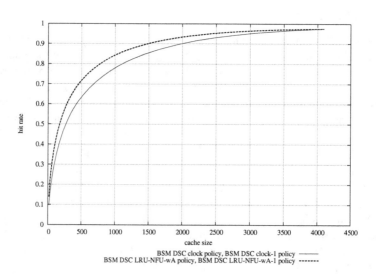

(b) Hit rate

Figure 2. Dependency on the cache size (English text, Levenshtein distance, $m = 50, n = 152089, k = 5, |Q_{\mathrm{NFA}}| = 291$).

When we compare the BSMCache to the DSCache performance, DSCache comes out as the winner. However, we must keep in mind how things work deep in these objects—that BSMCache utilizes a generic hash map, whereas DSCache is written especially for the purpose of DSC.

We want to point out another interesting observation. For a short pattern and a relatively high number of allowed errors, bit parallelism embodies worse running times than the original BSM. The reason is clear—in a such case, the automaton is still small and the bit-vector is small. As a consequence there are fast bit-vector operations and a fast BSM. On the other hand bit parallelism must still maintain a high number of tables (one for each error level) and execute many more bit-vector operations than the BSM.

7 Conclusion

A new approach to the simulation of general NFA has been introduced— the basic simulation method (BSM) utilizing a deterministic state cache (DSC) as a method aiming to speed up the original BSM. We have presented theoretical basis for the DSC.

We identified several important factors which influence the performance of DSC, and their relations. Namely they are the cache management overhead, the locality present in a particular automaton and in an input text. The time complexity for the BSM utilizing DSC is $\mathcal{O}(n(c_s\lceil\frac{|Q|}{w}\rceil + r_m(c_s|Q|\lceil\frac{|Q|}{w}\rceil + c_s\lceil\frac{|Q|}{w}\rceil)))$. The term $c_s\lceil\frac{|Q|}{w}\rceil$ represents the cache management overhead (the biggest part of it). The locality of the automaton and the locality of the input text input are aggregated into one term—miss rate r_m (or hit rate r_h consequently, $r_m + r_h = 1$). It is obvious that for any speed-up of the algorithm it is important to ensure that the term $r_m(c_s|Q|\lceil\frac{|Q|}{w}\rceil + c_s\lceil\frac{|Q|}{w}\rceil)$ is not significant (i.e., the cache look-up does not fail too often, and cache management requires a minimum time).

The experiments showed the real performance of DSC and its behavior related to the above-mentioned factors. Dependency on the cache management overhead can be reduced simply by using a specialized implementation of the whole cache structure as in DSCache, which has very persuasive results—in certain cases the hit rate fell almost to zero, However, the overhead of DSCache is almost insignificant, so the performance dropped only close to the performance of the original method (typically it was slightly worse). Unlike the DSCache, the cache management overhead of BSMCache, based on a generic hash table template from Google's SparseHash project, became insignificant with high hit rate (typically 0.8 and higher), otherwise the performance of this implementation was very poor.

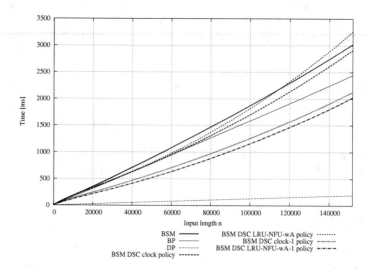

(a) run time

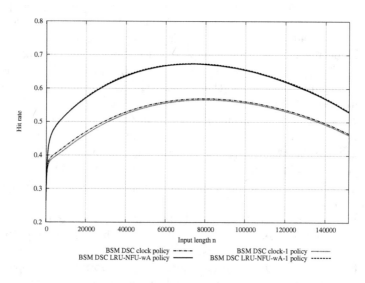

(b) hit rate

Figure 3. Input length n. English text, Levenshtein distance, Σ version, $\varepsilon CLOSURE$ preprocessed, $m = 15$, $k = 5$, $c_s = 128$.

The second factor is a locality contained in the automaton itself. In pattern matching NFAs we can observe a dependency on the number of allowed errors k which causes decreasing locality with growing k in the resulting automaton. We can also observe configurations with active states deeper in the automaton being used only once for a long period and therefore being purged from the cache before used again. Although DSCache still performed only slightly worse than BSM for automata with lack of locality, we rather choose another method for such automata. We have to keep in mind that in such a case the locality cannot be improved by using a bigger cache. Such an approach only brings the deterministic state cache closer to the determinization algorithm. Thus it inherits its characteristics, mainly a huge slow-down.

The last factor is the locality contained in the input text—mainly the dependency on the length of the input text. For short texts, the cache cannot even fill up, so a new set of active states must be computed by the BSM algorithm. In a contrast, cache purging and misses occur more often for long texts, so new set of active states must be computed by the basic simulation method algorithm again. However, in this case the hit rate can be improved by using a bigger cache and consequently speed-up can be observed.

When discussing the input text, we have to mention different performance for texts with a different character. Firstly, the character of the text typically specifies the alphabet that is used. Secondly, it influences the probabilities of the sequences of symbols, and so it influences the locality. However, in this case we are at the edge between locality in the automaton and locality in the text, and unfortunately this kind of locality cannot be improved by using a bigger cache.

The deterministic state cache approach has proved that in a wide range of cases speed-up can be achieved. In pattern matching, where NFAs have some regular structure of transitions, we have better methods (BP and DP) to use. However, this method is very important in other areas where we require simulation of general NFAs, so neither BP nor DP can be utilized. However, the above mentioned conditions still have to be kept in mind and most likely exhaustive experiments should be performed on these NFAs.

Acknowledgement

We thank the anonymous referees for offering valuable comments that have contributed to the quality of this paper.

BIBLIOGRAPHY

[1] Google SparseHash project. http://code.google.com/p/google-sparsehash/.

[2] A. V. Aho. Pattern matching in strings. In R. Book, editor, *Formal Language Theory: Perspectives and Open Problems*, pages 325–347. Academic Press, London, U. K., 1980.

[3] R. A. Baeza-Yates and G. H. Gonnet. A new approach to text searching. *Commun. ACM*, 35(10):74–82, 1992.

[4] R. A. Baeza-Yates and G. H. Gonnet. Fast string matching with mismatches. *Inf. Comput.*, 108(2):187–199, 1994.

[5] T. Bell. The Canterbury Corpus. http://corpus.canterbury.ac.nz/.

[6] B. Dömölki. An algorithm for syntactical analysis. *Computational Linguistics*, 3:29–46, 1964. Hungarian Academy of Science, Budapest.

[7] J. Holub. *Simulation of Nondeterministic Finite Automata in Pattern Matching*. Ph.D. Thesis, Czech Technical University in Prague, Czech Republic, February 2000.

[8] J. Holub. Bit parallelism—NFA simulation. In B.W. Watson and D. Wood, editors, *Implementation and Application of Automata*, number 2494 in Lecture Notes in Computer Science, pages 149–160. Springer-Verlag, Heidelberg, 2002.

[9] J. Holub. Dynamic programming for reduced NFAs for approximate string and sequence matching. *Kybernetika*, 38(1):81–90, 2002.

[10] J. Holub. Dynamic programming—NFA simulation. In J.-M. Champarnaud and D. Maurel, editors, *Implementation and Application of Automata*, number 2608 in Lecture Notes in Computer Science, pages 295–300. Springer-Verlag, Heidelberg, 2003.

[11] J. Holub and P. Špiller. Practical experiments with NFA simulation. In L. Cleophas and B. W. Watson, editors, *Proceedings of the Eindhoven FASTAR Days 2004*, pages 73–95. TU Eindhoven, The Netherlands, 2004.

[12] J. E. Hopcroft and J. D. Ullman. *Introduction to automata, languages and computations*. Addison-Wesley, Reading, MA, 1979.

[13] D. C. Kozen. *Automata and Computability*. Springer-Verlag, Berlin, 1997.

[14] S. Kurtz. *Fundamental Algorithms for a Declarative Pattern Matching System*. PhD thesis, Technische Fakultät, Universität Bielefeld, September 1995. Available as Report 95-03.

[15] G. Navarro. A partial deterministic automaton for approximate string matching. In R. Baeza-Yates, editor, *Proceedings of the 4th South American Workshop on String Processing*, pages 95–111, Valparaiso, Chile, 1997. Carleton University Press.

[16] P. Norvig. Techniques for automatic memoization with applications to context-free parsing. *Comput. Linguist.*, 17(1):91–98, 1991.

[17] P. H. Sellers. The theory and computation of evolutionary distances: Pattern recognition. *J. Algorithms*, 1(4):359–373, 1980.

[18] R. K. Shyamasundar. A simple string matching algorithm. technical report, Tata Institute of Fundamental Research, India, 1976. 9 pages.

[19] E. Ukkonen. Finding approximate patterns in strings. *J. Algorithms*, 6(1–3):132–137, 1985.

[20] S. Wu and U. Manber. Fast text searching allowing errors. *Commun. ACM*, 35(10):83–91, 1992.

Jan Holub
Department of Computer Science and Engineering
Czech Technical University in Prague
Karlovo nám. 13, 121 35, Prague 2, Czech Republic
Email: holub@fel.cvut.cz

Tomáš Kadlec
Department of Computer Science and Engineering

Czech Technical University in Prague
Karlovo nám. 13, 121 35, Prague 2, Czech Republic
Email: kadlet2@fel.cvut.cz

Repetitions and Factor Automata[1]

Bořivoj Melichar and Marek Hanuš

ABSTRACT. A way to find repetitions of factors in a given text is shown. Models for finding exact repetitions in one string and in a finite set of strings are introduced. It is shown that d-subsets created during determinisation of nondeterministic factor automata contain all information concerning repetitions of factors. The principle of the analysis of d-subsets is then used for finding repetitions.

1 Introduction

Let a text $T = t_1 t_2 \ldots t_n$ be given. Finding a repetition in text T can be defined as determining whether some substring (factor) repeats in a given text T. Furthermore, we can distinguish between exact and approximate repetitions. In some cases the problem of finding repetitions in a text concerns a specified factor.

We will deal with exact repetitions only in this paper. The goal of this work is to find a way to reveal all these repetitions in the given text that is easy to understand. The main idea is based on the use of factor automata which can be used to uniquely describe many problems.

One of the first attempts to solve problem of finding repetitions in a string was presented in [8]. It finds longest repeated factor (possibly overlapping) in $\mathcal{O}(n \log n)$ time, where n is length of the input text.

The first attempt to solve the problem of finding squares (two consecutive occurrences of some factor) in a string is in a master's thesis [10]. Time complexity of this approach is $\mathcal{O}(n \log n)$. A linear algorithm for searching a square was presented in [4]. It is based on construction of factor transducer.

Number of repetitions in a string can be at most $\mathcal{O}(n^2)$, where n is length of the string. To design algorithm that is running in time better than $\mathcal{O}(n^2)$, more efficient way of encoding repetitions is needed. First such algorithm was described in [3]. It is able to find all consecutively repeating factors in a string in $\mathcal{O}(n \log n)$ time. A linear time algorithm was presented in [9].

[1]This research has been partially supported by the Ministry of Education, Youth, and Sport of the Czech Republic under research program MSM 6840770014 and by the Czech Science Foundation as project No. 201/06/1039.

It finds all so-called runs. Run is maximal consecutively repeating factor non-extendable to the left nor right. This repetition can be fractional (i.e. its exponent need not to be an integer number).

The approach used in this paper is based on the construction of deterministic factor automata. It was described in a master's thesis [12] and internal technical report [11]. Time complexity is $\mathcal{O}(n+r)$, where n is length of the input text and r is number of all repetitions in the text.

After the overview of basic definitions in Section 2, the next two chapters are devoted to exact repetitions in one string and in a finite set of strings. Section 5 gives time and space complexities of our approach. Section 6 is conclusion.

2   Definitions

Basic notions from the theory of automata follow [1] and [7]. Basic notions from stringology are from [5]. The notion of a factor automaton was introduced in [2].

DEFINITION 1 (Deterministic finite automaton).

A *deterministic finite automaton* (*DFA*) is quintuple $M = (Q, A, \delta, q_0, F)$, where

Q is a finite set of states,

A is a finite input alphabet,

δ is a mapping from $Q \times A$ to Q,

$q_0 \in Q$ is an initial state,

$F \subset Q$ is the set of final states.

DEFINITION 2 (Nondeterministic finite automaton).

A *nondeterministic finite automaton* (*NFA*) is quintuple $M = (Q, A, \delta, q_0, F)$, where

Q is a finite set of states,

A is a finite input alphabet,

δ is a mapping from $Q \times A$ into the set of subsets of Q,

$q_0 \in Q$ is an initial state,

$F \subset Q$ is the set of final states.

DEFINITION 3 (Set of factors).

The set $Fact(x), x \in A^*$, is the set of all *substrings (factors)* of the string x:

$Fact(x) = \{y : x = uyv, u, v, x, y \in A^*\}$.

DEFINITION 4 (*d*-subset).

Let $M_1 = (Q_1, A, \delta_1, q_{01}, F_1)$ be a nondeterministic finite automaton. Let $M_2 = (Q_2, A, \delta_2, q_{02}, F_2)$ be the deterministic finite automaton equivalent to automaton M_1. Automaton M_2 is constructed using the standard determinisation algorithm based on subset construction. Every state $q \in Q_2$ corresponds to some subset d of Q_1. This subset will be called a d-*subset* (deterministic subset).

DEFINITION 5 (Exact repetition in one string).
Let T be a string, $T = a_1 a_2 \ldots a_n$ and $a_i = a_j, a_{i+1} = a_{j+1}, \ldots, a_{i+m} = a_{j+m}, i < j, m \geq 0$. String $x_2 = a_j a_{j+1} \ldots a_{j+m}$ is an *exact repetition* of the string $x_1 = a_i a_{i+1} \ldots a_{i+m}$. x_1 or x_2 are called repeating factors in text T.

DEFINITION 6 (Exact repetition in a set of strings).
Let S be a set of strings, $S = \{x_1, x_2, \ldots, x_{|S|}\}$ and $x_{pi} = x_{qj}, x_{pi+1} = x_{qj+1}, \ldots, x_{pi+m} = x_{qj+m}, p \neq q$ or $p = q$ and $i < j, m \geq 0$. String $x_{qj} x_{qj+1} \ldots x_{qj+m}$ is an *exact repetition* of string $x_{pi} x_{pi+1} \ldots x_{pi+m}$.

DEFINITION 7 (Type of repetition).
Let $x_2 = a_j a_{j+1} \ldots a_{j+m}$ be an exact repetition of $x_1 = a_i a_{i+1} \ldots a_{i+m}$, $i < j$, in one string.
Then if $j - i < m$ the repetition is with an *overlap* (O),
if $j - i = m$ the repetition is a *square* (S),
if $j - i > m$ the repetition is with a *gap* (G).

3 Exact repetitions in one string

In this section we will introduce how to use a factor automaton for finding exact repetitions in one string. The main idea is based on constructing a deterministic factor automaton. First, a nondeterministic factor automaton for a given string is created. The next step is to construct the equivalent deterministic factor automaton. During this process, d-subsets are stored. The repetitions that we are looking for are obtained by analysing these d-subsets. The whole process is formalized in Algorithm 1.

Algorithm 1
Computation of repetitions in one string.
Input: String $T = a_1 a_2 \ldots a_n$.
Output: Deterministic factor automaton M_D accepting $Fact(T)$ and d-subsets for all states of M_D.
Method:

1. Construct the nondeterministic factor automaton M_N accepting the set $Fact(T)$:

   (a) Construct the finite automaton M accepting the string $T = a_1 a_2 \ldots a_n$ and all its prefixes:
   $M = (\{q_0, q_1, q_2, \ldots, q_n\}, A, \delta, q_0, \{q_0, q_1, \ldots, q_n\})$,
   where $\delta(q_i, a_{i+1}) = q_{i+1}$ for all $i \in \langle 0, n-1 \rangle$.

(b) Construct finite automaton M_ε from the automaton M by inserting the ε-transitions:
$$\delta(q_0, \varepsilon) = \{q_1, q_2, \ldots, q_{n-1}, q_n\}.$$

(c) Replace all ε-transitions by non-ε-transitions. The resulting automaton is M_N.

2. Construct deterministic factor automaton M_D equivalent to automaton M_N and store the d-subsets during this construction.

3. Analyze the d-subsets to compute the repetitions. □

The factor automaton M_ε constructed by Algorithm 1 has, after step 1b, the transition diagram depicted in Figure 1. After step 1c of Algorithm 1, factor automaton M_N has the transition diagram depicted in Figure 2.

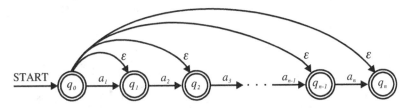

Figure 1. Transition diagram of factor automaton M_ε with ε-transitions constructed in step 1b of Algorithm 1.

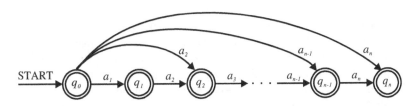

Figure 2. Transition diagram of factor automaton M_N after removal of the ε-transitions in step 1c of Algorithm 1.

The next example shows the construction of the deterministic factor automaton and the analysis of the d-subsets.

Let us make a note concerning labelling: Labels used as the names of states are selected in order to indicate positions in the string. This labelling will be useful later.

Example 1

Let us use the text $T = ababa$. First, we construct the nondeterministic factor automaton $M_\varepsilon(ababa) = (Q_\varepsilon, A, \delta_\varepsilon, 0, Q_\varepsilon)$ with ε-transitions. Its transition diagram is depicted in Figure 3.

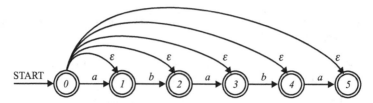

Figure 3. Transition diagram of the factor automaton $M_\varepsilon(ababa)$ from Example 1.

Then the ε-transitions are removed. The resulting nondeterministic factor automaton $M_N(ababa) = (Q_N, A, \delta_N, 0, Q_N)$ is depicted in Figure 4 and its transition table is Table 1.

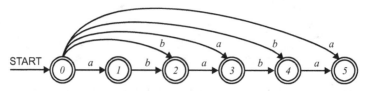

Figure 4. Transition diagram of the nondeterministic factor automaton $M_N(ababa)$ from Example 1.

In the next step, we construct the equivalent deterministic factor automaton $M_D(ababa) = (Q_D, A, \delta_D, 0, Q_D)$. During this operation the created d-subsets are stored. We assume, taking into account the labelling of the states of the nondeterministic factor automaton, that the d-subsets are ordered in the natural way. The extended transition table (with ordered d-subsets) of the deterministic factor automaton $M_D(ababa)$ is shown in Table 2. A transition diagram of M_D is depicted in Figure 5.

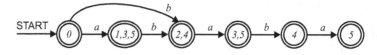

Figure 5. Transition diagram of the deterministic factor automaton $M_D(ababa)$ from Example 1.

Now we start the analysis of the resulting d-subsets: The d-subset $d(D_1) = \{1, 3, 5\}$ shows that factor a repeats at positions $1, 3$ and 5 of the

State	a	b
0	1, 3, 5	2, 4
1		2
2	3	
3		4
4	5	
5		

Table 1.   Transition table of the nondeterministic factor automaton $M_N(ababa)$ from Example 1.

State	d-subset	a	b
D_0	0	1, 3, 5	2, 4
D_1	1, 3, 5		2, 4
D_2	2, 4	3, 5	
D_3	3, 5		4
D_4	4	5	
D_5	5		

Table 2. Transition table of the automaton $M_D(ababa)$ from Example 1.

given string, and its length is one. The d-subset $d(D_2) = \{2, 4\}$ shows that factor ab repeats, and its occurrence in the string ends at positions 2 and 4 and its length is two. Moreover, suffix b of this factor also repeats at the same positions as factor ab. The d-subset $d(D_3) = \{3, 5\}$ shows that factor aba repeats, and its occurrence in the string ends at positions 3 and 5 and its length is three. Its suffix ba also repeats at the same positions. Suffix a of factor aba also repeats at positions 3 and 5, but we have already obtained this information during analysis of the d-subset $d(D_1) = \{1, 3, 5\}$. Analysis of the d-subsets having only single states provides no further information on repeating factors.

A summary of these observations is collected in a *repetition table*. The repetition table contains one row for each d-subset. It contains the d-subset, the repeating factor, and a list of repetitions. The list of repetitions indicates the position of the repeating factor and the type of repetition.   □

DEFINITION 8.

Let T be a string. The repetition table for T contains the following items:

1. d-subset,

2. corresponding factor,

3. list of repetitions of the factor containing elements of the form (i, X_i),
   where i is the position of the factor in string T,
            X_i is the type of repetition:
                    F - first occurrence of the factor,
                    O - repetition with overlapping,
                    S - repetition as a square,
                    G - repetition with a gap.

The repetition table for the string $T = ababa$ from Example 1 is shown in Table 3.

d-subset	Factor	List of Repetitions
$1, 3, 5$	a	$(1, F), (3, G), (5, G)$
$2, 4$	ab	$(2, F), (4, S)$
$2, 4$	b	$(2, F), (4, G)$
$3, 5$	aba	$(3, F), (5, O)$
$3, 5$	ba	$(3, F), (5, S)$

Table 3. Repetition table of $ababa$.

The construction of the repetition table is based on the following observations illustrated in Figure 6, and Lemmata 9 and 10 show its correctness.

LEMMA 9. *Let T be a string and $M_D(T)$ be the deterministic factor automaton for T with states labelled by the corresponding d-subsets. If factor $u = a_1 a_2 \ldots a_m$, $m \geq 1$, repeats in string T and its occurrences start at positions $x + 1$ and $y + 1$, $x \neq y$ then there exists a d-subset in $M_D(T)$ containing the pair $\{x + m, y + m\}$.*

Proof. Let $M_N(T) = (Q_N, A, \delta_N, q_0, Q_N)$ be the nondeterministic factor automaton for T and let $u = a_1 a_2 \ldots a_m$ be the factor starting at positions $x + 1$ and $y + 1$ in T, $x \neq y$. Then there are transitions in $M_N(T)$ from state 0 to states $x + 1$ and $y + 1$ for symbol a_1, $(\delta_N(0, a_1)$ contains $x + 1$ and $y + 1)$. It follows from the construction of $M_N(T)$ that:

$$\delta_N(x + 1, a_2) = \{x + 2\}, \qquad\qquad \delta_N(y + 1, a_2) = \{y + 2\},$$
$$\delta_N(x + 2, a_3) = \{x + 3\}, \qquad\qquad \delta_N(y + 2, a_3) = \{y + 3\},$$
$$\vdots \qquad\qquad\qquad\qquad\qquad\qquad \vdots$$
$$\delta_N(x + m - 1, a_m) = \{x + m\}, \qquad \delta_N(y + m - 1, a_m) = \{y + m\}.$$

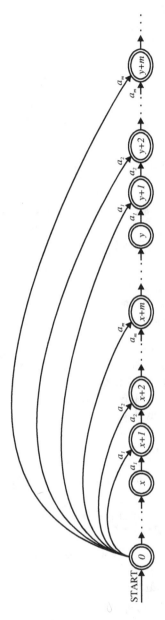

Figure 6. Repeated factor $u = a_1 a_2 \ldots a_m$ in $M_N(T)$.

The deterministic factor automaton $M_D(T) = (Q_D, A, \delta_D, D_0, Q_D)$ then contains the states $D_0, D_1, D_2, \ldots D_m$ having the following property:

$$\delta_D(D_0, a_1) = D_1, \qquad \{x+1, y+1\} \subset D_1,$$
$$\delta_D(D_1, a_2) = D_2, \qquad \{x+2, y+2\} \subset D_2,$$
$$\vdots \qquad\qquad\qquad \vdots$$
$$\delta_D(D_{m-1}, a_m) = D_m, \qquad \{x+m, y+m\} \subset D_m.$$

We can conclude that the d-subset D_m contains the pair $\{x+m, y+m\}$. ∎

LEMMA 10. *Let T be a string and let $M_D(T)$ be the deterministic factor automaton for T with the states labelled by the corresponding d-subsets. If a d-subset D_m contains two elements $x+m$ and $y+m$ then there exists the factor $u = a_1 a_2 \ldots a_m$, $m \geq 1$, starting at both positions x and y in string T.*

Proof. Let $M_N(T)$ be the nondeterministic factor automaton for T. If a d-subset D_m contains elements from $\{x+m, y+m\}$ then it holds for δ_N of $M_N(T)$:
$\{x+m, y+m\} \subset \delta_N(0, a_m)$, and
$\delta_N(x+m-1, a_m) = \{x+m\}$,
$\delta_N(y+m-1, a_m) = \{y+m\}$ for some $a_m \in A$.
Then d-subset D_{m-1} such that $\delta_D(D_{m-1}, a_m) = D_m$ must contain
$x+m-1, y+m-1$ such that $\{x+m-1, y+m-1\} \subset \delta_N(0, a_{m-1})$,
$\delta_N(x+m-2, a_{m-1}) = \{x+m-1\}$,
$\delta_N(y+m-2, a_{m-1}) = \{y+m-1\}$
and for the same reason d-subset D_1 must contain $x+1, y+1$ such that
$\{x+1, y+1\} \subset \delta_N(0, a_1)$ and $\delta_N(x, a_1) = \{x+1\}$, $\delta_N(y, a_1) = \{y+1\}$.

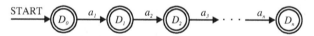

Figure 7. Repeated factor $u = a_1 a_2 \ldots a_m$ in $M_D(T)$.

Then there exists the sequence of transitions in $M_D(T)$ (see Figure 7):

$$(D_0, a_1 a_2 \ldots a_m) \quad \vdash \quad (D_1, a_2 \ldots a_m)$$
$$\vdash \quad (D_2, a_3 \ldots a_m)$$
$$\vdots$$
$$\vdash \quad (D_{m-1}, a_m)$$

$$\vdash (D_m, \varepsilon),$$

where

$$\{x+1, y+1\} \subset D_1,$$
$$\vdots$$
$$\{x+m, y+m\} \subset D_m.$$

This sequence of transitions corresponds to two different sequences of transitions in $M_N(T)$ going through state $x+1$:

$$
\begin{array}{ll}
(0, a_1 a_2 \ldots a_m) & \vdash (x+1, a_2 \ldots a_m) \\
& \vdash (x+2, a_3 \ldots a_m) \\
& \vdots \\
& \vdash (x+m-1, a_m) \\
& \vdash (x+m, \varepsilon), \\
(x, a_1 a_2 \ldots a_m) & \vdash (x+1, a_2 \ldots a_m) \\
& \vdash (x+2, a_3 \ldots a_m) \\
& \vdots \\
& \vdash (x+m-1, a_m) \\
& \vdash (x+m, \varepsilon).
\end{array}
$$

Similarly two sequences of transitions go through state $y+1$:

$$
\begin{array}{ll}
(0, a_1 a_2 \ldots a_m) & \vdash (y+1, a_2 \ldots a_m) \\
& \vdash (y+2, a_3 \ldots a_m) \\
& \vdots \\
& \vdash (y+m-1, a_m) \\
& \vdash (y+m, \varepsilon), \\
(y, a_1 a_2 \ldots a_m) & \vdash (y+1, a_2 \ldots a_m) \\
& \vdash (y+2, a_3 \ldots a_m) \\
& \vdots \\
& \vdash (y+m-1, a_m) \\
& \vdash (y+m, \varepsilon).
\end{array}
$$

It follows from this that the factor $u = a_1 a_2 \ldots a_m$ is present twice in string T at different positions $x+1, y+1$. ∎

The following Lemma is a simple consequence of Lemma 10.

LEMMA 11. *Let u be a repeating factor in string T. Then all factors of u are also repeating factors in T.*

4   Exact repetitions in a finite set of strings

The idea of using a factor automaton for finding exact repetitions in one string can also be used for finding exact repetitions in a finite set of strings. In the next example we show the construction of a factor automaton and an analysis of the d-subsets created during this construction.

Example 2

Let us construct the factor automaton for the set of strings $S = \{abab, abba\}$. First, we construct the factor automata $M_{1\varepsilon}$ and $M_{2\varepsilon}$ for both strings in S. Their transition diagrams are depicted in Figures 8 and 9, respectively.

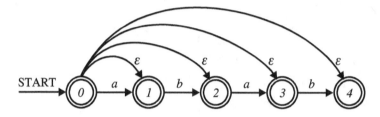

Figure 8. Transition diagram of factor automaton $M_{1\varepsilon}$ accepting the set $Fact(abab)$ from Example 2.

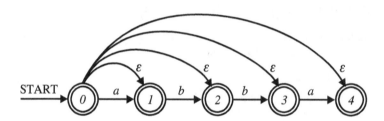

Figure 9. Transition diagram of factor automaton $M_{2\varepsilon}$ accepting the set $Fact(abba)$ from Example 2.

In the second step automaton M_ε accepting the language $L(M_\varepsilon) = Fact(abab) \cup Fact(abba)$ is constructed. Its transition diagram is depicted in Figure 10.

In the third step automaton M_N is constructed by removing the ε-transitions from automaton M_ε. Its transition diagram is depicted in Figure 11, its transition table is Table 4.

The last step is to construct the deterministic factor automaton M_D. Its transition table is shown in Table 5. The transition diagram of the resulting deterministic factor automaton M_D is depicted in Figure 12.

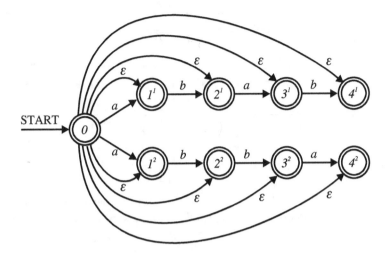

Figure 10. Transition diagram of factor automaton M_ε accepting the set $Fact(abab) \cup Fact(abba)$ from Example 2.

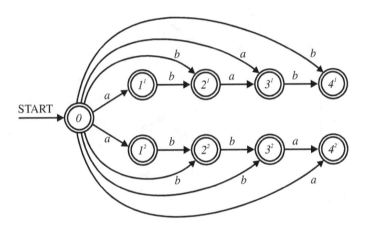

Figure 11. Transition diagram of nondeterministic factor automaton M_N accepting the set $Fact(abab) \cup Fact(abba)$ from Example 2.

Now we make an analysis of the d-subsets of the resulting automaton M_D. The result of this analysis is the repetition table shown in Table 6 for the set $S = \{abab, abba\}$.   □

DEFINITION 12.
    Let S be a set of strings $S = \{x_1, x_2, \ldots, x_{|S|}\}$. The repetition table for

State	a	b
0	$1^1,1^2,3^1,4^2$	$2^1,2^2,3^2,4^1$
1^1		2^1
1^2		2^2
2^1	3^1	
2^2		3^2
3^1		4^1
3^2	4^2	
4^1		
4^2		

Table 4. Transition table of the nondeterministic factor automaton M_N accepting the set $Fact(abab) \cup Fact(abba)$ from Example 2.

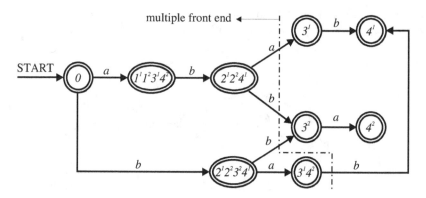

Figure 12. Transition diagram of the deterministic factor automaton M_D accepting the set $Fact(abab) \cup Fact(abba)$ from Example 2.

S contains the following items:

1. d-subset,

2. corresponding factor,

3. the list of repetitions of the factor containing elements of the form (i, j, X_{ij}), where

d-subset	a	b
0	$1^1,1^2,3^1,4^2$	$2^1,2^2,3^2,4^1$
$1^1,1^2,3^1,4^2$		$2^1,2^2,4^1$
$2^1,2^2,3^2,4^1$	$3^1,4^2$	3^2
$2^1,2^2,4^1$	3^1	3^2
$3^1,4^2$		4^1
3^1		4^1
3^2	4^2	
4^1		
4^2		

Table 5. Transition table of the deterministic factor automaton M_D from Example 2.

d-subset	Factor	List of Repetitions
$1^1 1^2 3^1 4^2$	a	$(1,1,F),(2,1,F),(1,3,G),(2,4,G)$
$2^1 2^2 4^1$	ab	$(1,2,F),(2,2,F),(1,4,S)$
$2^1 2^2 3^2 4^1$	b	$(1,2,F),(2,2,F),(2,3,S),(1,4,G)$
$3^1 4^2$	ba	$(1,3,F),(2,4,F)$

Table 6. Repetition table for the set $S = \{abab, abba\}$ from Example 2.

i is the index of the string in S,
j is the position in string x_i,
X_{ij} is the type of repetition:
$\quad$ F - the first occurrence of the factor in string x_i,
$\quad$ O - repetition of the factor in x_i with overlapping,
$\quad$ S - repetition as a square in x_i,
$\quad$ G - repetition with a gap in x_i.

5 Time and space complexity

Time and space complexity of our approach will be determined in this section.

PROPOSITION 13. *Let n be the length of the pattern and let r be a number of repetitions (i.e. size of the repetition table). Time and space complexities of Algorithm 1 are $\mathcal{O}(n+r)$.*

Proof. Nondeterministic factor automaton M_N can be constructed in $\mathcal{O}(n)$ time. Determinisation of this automaton is dependent on implementation of the d-subsets. In case of exact repetitions, a simple observations can be made: The first row of the transition table of the nondeterministic factor automaton M_N is the same as the first row of the transition table of the deterministic factor automaton M_D. Therefore, it can be copied in $\mathcal{O}(n)$ time. States corresponding to the other rows of the transition table of M_N are deterministic. Moreover, columns of the table are sorted if states of M_N are labelled properly (see Figure 4 and Table 1, Figure 11 and Table 4). d-subsets can be implemented as vectors then – no sets are needed.

As proved in [6], factor automaton for text of length n has at most $2n$ states. This means that during process of deteminisation at most $2n$ unique d-subsets will be created. Sum of sizes of all these d-subsets is at most $n + 1 + r$ since d-subsets greater than one contain at most r elements in total (number of repetitions) and there is at most $n + 1$ d-subsets of size one. d-subsets of size one need not to be constructed (except initial state) during determinisation – no d-subset of size greater than one can be created as a successor of any d-subset of size one (see Figure 12). Therefore sum of sizes of all d-subsets created during determinisation is at most $\mathcal{O}(r)$. For every element of these d-subsets except initial state, one or no transition in M_N exists. This transition has to be added to the d-subset being created during determinisation. This can be done in constant time since columns of the transition table of M_N are sorted – we obtain sorted d-subsets only by appending the new element to the end. Sorted d-subsets are important to find whether newly created d-subsets are unique in $\mathcal{O}(r)$ time in total. The overall complexity of determinisation is then $\mathcal{O}(n + r)$.

Analysis of d-subsets can be performed by traversing the deterministic factor automaton M_D from the initial state. This can be done in $\mathcal{O}(r)$ time since a new row is added into repetition table in every step. The overall time complexity of the algorithm is then $\mathcal{O}(n + r)$.

As for the space complexity, the biggest structures needed are transition tables of the deterministic and nondeterministic factor automata. Transition table of the nondeteministic factor automaton M_N can be stored in $\mathcal{O}(n)$ space since initial state has n outgoing transitions and all other states has one or no outgoing transition. Transition table of the deteministic factor automaton M_D can be stored in $\mathcal{O}(n + r)$ space if sparse representation is used. The overall space complexity is then $\mathcal{O}(n + r)$. ∎

Proposition 13 holds also for exact repetitions in a finite set of strings. Then n is sum of lengths of all patterns in the set.

Lower bound on the number of repetitions r is zero for strings containing every symbol just once. Upper bound is $\frac{1}{2}(n^2 + n - 2)$ for string a^n, $a \in A$,

$n \in \mathbb{N}$.

6   Conclusions

We have shown uniform and simple models for finding exact repetitions in a text. All models are based on the analysis of d-subsets created during determinisation of different types of nondeterministic factor automata. This leads to a very simple and straightforward solution of repetition problems.

A topic for further study is the search for a connection between the models shown here and existing algorithms for repetition problems. This may lead to an understanding of the way to simulate general models in different cases.

Acknowledgements We are grateful to Olga Vrtišková for preparing the text and drawing the pictures. Our thanks go to Robin Healey for revising the English text.

BIBLIOGRAPHY

[1] A. V. Aho and J. D. Ullman. *The theory of parsing, translation, and compiling.* Prentice-Hall, Upper Saddle River, NJ, USA, 1972.

[2] A. Blumer, J. Blumer, D. Haussler, R. McConnell, and A. Ehrenfeucht. Complete inverted files for efficient text retrieval and analysis. *Journal of the ACM*, 34(3):578–595, 1987.

[3] M. Crochemore. An optimal algorithm for computing the repetitions in a word. *Information Processing Letters*, 12(5):244–250, 1981.

[4] M. Crochemore. Transducers and repetitions. *Theoretical Computer Science*, 45(1):63–86, 1986.

[5] M. Crochemore and W. Rytter. *Text algorithms.* Oxford University Press, Inc., New York, NY, USA, 1994.

[6] M. Crochemore and W. Rytter. *Jewels of stringology.* World Scientific Publishing Company, Singapore, 2002.

[7] J. E. Hopcroft and J. D. Ullman. *Introduction to automata, languages and computations.* Addison-Wesley, Reading, MA, USA, 1979.

[8] R. M. Karp, R. E. Miller, and A. L. Rosenberg. Rapid identification of repeated patterns in strings, trees and arrays. In *STOC '72: Proceedings of the fourth annual ACM symposium on Theory of computing*, pages 125–136, New York, NY, USA, 1972. ACM.

[9] R. Kolpakov and G. Kucherov. Finding maximal repetitions in a word in linear time. In *Proceedings of the 1999 Symposium on Foundations of Computer Science*, pages 596–604. IEEE Computer Society, 1999.

[10] M. Main. An $\mathcal{O}(n \log n)$ algorithm for finding repetition in a string. Master's thesis, Washington State University, 1979.

[11] B. Melichar. Repetitions in text and finite automata. In *Proceedings of the Eindhoven FASTAR Days 2004*, Eindhoven, 2004. Department of Mathematics and Computer Science, Technical University Eindhoven. CS-Report04-40.

[12] B. Melichar(jr.). Repetitions in strings. Master's thesis, Czech Technical University in Prague, 2002.

Bořivoj Melichar
Prague Stringology Club
Department of Computer Science & Engineering
Czech Technical University in Prague
Karlovo nám. 13, 121 35 Prague 2, Czech Republic
Email: melichar@fel.cvut.cz
http://www.stringology.org/athens

Marek Hanuš
Prague Stringology Club
Department of Computer Science & Engineering
Czech Technical University in Prague
Karlovo nám. 13, 121 35 Prague 2, Czech Republic
Email: hanusm2@fel.cvut.cz
http://www.stringology.org/athens

String Comparison by Transposition Networks

PETER KRUSCHE AND ALEXANDER TISKIN

ABSTRACT. Computing string or sequence alignments is a classical method of comparing strings and has applications in many areas of computing, such as signal processing and bioinformatics. Semi-local string alignment is a recent generalisation of this method, in which the alignment of a given string and all substrings of another string are computed simultaneously at no additional asymptotic cost. In this paper, we show that there is a close connection between semi-local string alignment and a certain class of traditional comparison networks known as transposition networks. The transposition network approach can be used to represent different string comparison algorithms in a unified form, and in some cases provides generalisations or improvements on existing algorithms. This approach allows us to obtain new algorithms for sparse semi-local string comparison and for comparison of highly similar and highly dissimilar strings, as well as of run-length compressed strings. We conclude that the transposition network method is a very general and flexible way of understanding and improving different string comparison algorithms, as well as their efficient implementation.

1 Introduction

In this paper we look at the classical problem of computing the cost of string (or sequence) alignments, particularly the longest common subsequence and edit distance problems. Since this problem was originally proposed [1], a multitude of algorithms have been found to compute edit distances or equivalently longest common subsequences of two input strings (see e.g. [2, 3] for an overview). An interesting extension to string alignment is semi-local string comparison. In this problem, we are interested in computing longest common subsequence lengths for one string and all substrings of the other string. Schmidt [4] proposed an algorithm for computing all longest paths in grid dags which was applied to string-substring longest common subsequence (LCS) computation by Alves et al. [5], who found an algorithm which runs in $O(n^2)$ time. Tiskin [6] developed further understanding of the algorithm and its data structures, obtaining a subquadratic time algorithm for semi-local string comparison including string-substring and prefix-suffix LCS computation. Semi-local string comparison is useful as an intermediate step towards fully-local string comparison, in which all pairs of substrings of the input strings are compared. A straightforward application is e.g. computing the LCS efficiently in a sliding window (a slightly simpler version of this problem was studied in [7]). Semi-local string comparison is also a useful tool for obtaining efficient parallel algorithms for LCS computation [8, 9, 10]. A summary of other algorithmic applications is given in [11].

In this paper, we develop a new interpretation of standard and semi-local LCS algorithms, based on a certain class of traditional comparison networks known as transposition networks. This approach allows us to obtain new algorithms for sparse semi-local string comparison and for comparison of highly similar and highly dissimilar strings, as well as semi-local comparison of run-length compressed strings. The remainder of this paper is structured as follows. We introduce the necessary concepts of string comparison in Section 2, and describe the transposition network method in Section 3. We then show new algorithms for sparse semi-local string comparison in Section 4, show how to compare run-length compressed strings semi-locally in Section 5, and discuss comparing highly similar or highly dissimilar strings in Section 6.

2   String comparison

Let $x = x_1x_2\ldots x_m$ and $y = y_1y_2\ldots y_n$ be two strings over an alphabet Σ of size σ. We distinguish between consecutive *substrings* of a string x which can be obtained by removing zero or more characters from the beginning and/or the end of x, and *subsequences* which can be obtained by deleting zero or more characters in any position. The *longest common subsequence* (LCS) of two strings is the longest string that is a subsequence of both input strings, its length p (the LLCS) is a measure for the similarity of the two strings. Throughout this paper we will denote the set of integers $\{i, i+1, \ldots, j\}$ by $[i : j]$, and the set $\{i + \frac{1}{2}, i + \frac{3}{2}, \ldots, j - \frac{1}{2}\}$ of odd half-integers by $\langle i : j \rangle$. We will further mark odd half-integer variables using a $\hat{\ }$ symbol.

DEFINITION 1. Let the *alignment dag* be defined by a set of vertices $v_{i,j}$ with $i \in [0 : m]$ and $j \in [0 : n]$ and edges as follows. We have horizontal and vertical edges $v_{i,j-1} \to v_{i,j}$ and $v_{i-1,j} \to v_{i,j}$ of weight 0. Further, we introduce diagonal edges $v_{i-1,j-1} \to v_{i,j}$ of weight 1 which are present only if $x_i = y_j$.

Longest common subsequences of a substring $x_ix_{i+1}\ldots x_j$ and y corre-

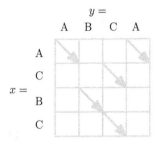

$$y =$$

A B C A

$$x =$$

A
C
B
C

$$L = \begin{pmatrix} \boxed{1} & 1 & 1 & 1 \\ 1 & 1 & \boxed{2} & 2 \\ 1 & \boxed{2} & \textcircled{2} & 2 \\ 1 & 2 & \boxed{3} & 3 \end{pmatrix}$$

(a) input strings and alignment
dag

(b) prefix-prefix LCS lengths
and contours

Figure 1. Example showing the alignment dag and the matrix L of prefix-prefix LCS lengths.

spond to longest paths in this graph from $v_{i-1,0}$ to $v_{j,m}$.

DEFINITION 2. We define the *extended alignment dag* as the infinite horizontal extension of the alignment dag, having vertices $v_{i,j}$ as above, but allowing $j \in [-\infty : \infty]$, adding corresponding horizontal and vertical edges as above for all additional vertices, and further including diagonal edges $v_{i-1,j-1} \rightarrow v_{i,j}$ of weight 1 for all $j < 1$ and $j > n$.

For many applications, the LCS itself is of lesser interest than its length. Looking at the LLCS for different substrings, including prefixes or suffixes of the input strings, exposes not only their global similarity, but also locations of high or low similarity. For example, the standard dynamic programming LCS algorithm compares all prefixes of one string to all prefixes of the other string [12] and stores their LCS lengths in the dynamic programming matrix $L(i,j) = \text{LLCS}(x_1 \dots x_i, y_1 \dots y_j)$. Semi-local string comparison [6] is an alternative to this standard string alignment method. Solutions to the semi-local LCS problem are given by a *highest-score matrix* which we define as follows.

DEFINITION 3. In a *highest-score matrix* A, each entry $A(i,j)$ is defined as the LLCS of x and substring $y_i \dots y_j$.

The definition of highest-score matrices can also be extended to include the LLCS of all *prefixes* $x_1 x_2 \dots x_i$ and all *suffixes* $y_j \dots y_n$, or the LLCS of all suffixes $x_i \dots x_m$ and all prefixes $y_1 \dots y_j$.

Since the values of $A(i,j)$ for different i and j are strongly correlated, it is possible to derive an implicit, space-efficient representation of matrix

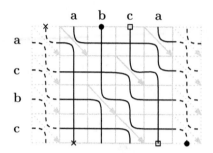
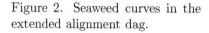

Figure 2. Seaweed curves in the extended alignment dag.

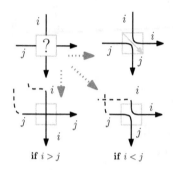

Figure 3. Illustration of the seaweed algorithm.

$A(i, j)$. This implicit representation of a semi-local highest-score matrix consists of a set of *critical points*.

DEFINITION 4. The critical points of a highest-score matrix A are defined as the set of odd half-integer pairs $(\hat{\imath}, \hat{\jmath})$ such that $A(\hat{\imath} + \frac{1}{2}, \hat{\jmath} - \frac{1}{2}) + 1 = A(\hat{\imath} - \frac{1}{2}, \hat{\jmath} - \frac{1}{2}) = A(\hat{\imath} + \frac{1}{2}, \hat{\jmath} + \frac{1}{2}) = A(\hat{\imath} - \frac{1}{2}, \hat{\jmath} + \frac{1}{2})$.

Tiskin [6] showed that in order to represent a highest-score matrix for two strings of lengths m and n, exactly $m + n$ such critical points are necessary. Note that infinitely many critical points exist in the extended alignment dag. However, due to the structure of the extended alignment dag, only a *core* of $m + n$ critical points need to be stored. Each of the remaining *off-core* critical points can be computed in constant time.

THEOREM 5. *The highest-score matrix A can be represented implicitly using only $O(m + n)$ space by its core critical points. We have: $A(i, j) = j - i - |\{(\hat{\imath}, \hat{\jmath}) : (\hat{\imath}, \hat{\jmath})$ is a critical point, $i < \hat{\imath}$ and $\hat{\jmath} < j\}|$.*

Proof. See [6].   ∎

Theorem 5 is a direct consequence of the Monge properties [13] of highest-score matrices. This theorem is particularly useful as it was also shown possible to combine two highest-score matrices in subquadratic time using their implicit representation [6] in order to obtain the highest-score matrix corresponding to comparing one string and a concatenation of two other strings. In [9, 14, 10], these methods were applied to obtaining efficient parallel algorithms for the LCS problem.

The set of critical points can be obtained using the *seaweed algorithm* (by Alves et al. [5], based on Schmidt [4], adapted by Tiskin [6]) which computes critical points incrementally for all prefixes of the input strings.

This dynamic programming procedure is graphically illustrated by tracing *seaweed curves* that start at odd half-integer positions between two adjacent vertices $v_{0,\hat{i}-\frac{1}{2}}$ and $v_{0,\hat{i}+\frac{1}{2}}$ in the top row of the extended alignment dag, and end between two adjacent vertices $v_{m,\hat{j}-\frac{1}{2}}$ and $v_{m,\hat{j}+\frac{1}{2}}$ in the bottom row (see Figure 2). Each critical point is computed as the pair of horizontal start and end coordinates of such a seaweed curve.

DEFINITION 6. Given the sequence $\{(\hat{i},\hat{j}_k) \; : \; k \in [1:m]\}$ where $(\hat{i},\hat{j}_k)$ is a critical point in the highest-score matrix of $x_1 \ldots x_k$ and y, a seaweed curve is obtained by connecting the sequence of points $\{(\hat{i},\hat{j}_1),(\hat{j}_1,\hat{j}_2),\ldots,(\hat{j}_{m-1},\hat{j}_m)\}$.

When drawing the (extended) alignment dag in the plane, its horizontal and vertical edges partition the plane into rectangular cells which, depending on the input strings, may contain a diagonal edge or not.

DEFINITION 7. For every pair of characters x_i and y_j we define a corresponding *cell* $(i - \frac{1}{2}, j - \frac{1}{2})$. Cells corresponding to a matching pair of characters will be called *match cells*, cells corresponding to mismatching characters or to cells only present in the extended alignment dag will be called *mismatch cells*.

Two seaweed curves enter every cell in the extended alignment dag, one at the left and one at the top. The seaweed curves proceed through the cell either downwards or rightwards. In the cell, the directions of these curves are interchanged either if there is a match $x_k = y_l$, or if the same pair of seaweed curves have already crossed. Otherwise, their directions remain unchanged and the curves cross. The seaweed algorithm is illustrated in Figure 3.

More efficient special case algorithms for the LCS problem can be obtained when parameterizing either by the number r of match cells, by the length p of the LCS, or by the edit distance. Previously, high-similarity string comparison has been considered in [15, 16, 17, 18, 19, 20, 21, 22]; all these papers give LCS algorithms for highly similar strings, running in time $O(ne)$, where e is either the edit distance between the strings (as in [17]), or a different closely related similarity measure. High-dissimilarity string comparison has been considered in [15, 21, 22]; the best running time for LCS on highly dissimilar strings is $O(np + n \log n)$. A good survey of parameterized string comparison algorithms is given by [2].

The basis of parameterized LCS computation for dissimilar strings is to determine the LCS of two strings as a longest *chain* of match cells $(i_1,j_1),(i_2,j_2),\ldots,(i_p,j_p)$ with $i_1 < i_2 < \ldots < i_p$ and $j_1 < j_2 < \ldots < j_p$. We define a partial order on the set of match cells by $(i_1,j_1) \prec (i_2,j_2)$ iff. $i_1 < i_2$ and $j_1 < j_2$; further, we say that (i_1,j_1) is dominated by (i_2,j_2).

Due to Dilworth's lemma [23], the minimum number of antichains (sets of pairwise incomparable elements) necessary to cover a partially ordered set is equal to the length of the longest chain. Therefore, the LCS of two strings can be obtained by computing a minimal antichain decomposition of the set of matches under the $\prec$ ordering. Consider chains ending at a match (i, j). If any longest such chain has length k, then this match is said to have *rank* k. If match (i, j) has rank k and for all other matches (i', j') of rank k either $i' \geq i$ and $j' < j$ or $j' \geq j$ and $i' < i$, then match (i, j) is called $(k$-)dominant.

The set of all dominant matches completely specifies the table of prefix-prefix LCS lengths $L(i, j) = LLCS(x_1 \ldots x_i, y_1 \ldots y_j)$. Let the contours of L be formed by the rows and columns of cells through which the values of L increase by one. A cell $(\hat{i}, \hat{j})$ belongs to a contour in L if $L(\hat{i} + \frac{1}{2}, \hat{j} + \frac{1}{2}) > L(\hat{i} - \frac{1}{2}, \hat{j} + \frac{1}{2})$, $L(\hat{i} + \frac{1}{2}, \hat{j} + \frac{1}{2}) > L(\hat{i} + \frac{1}{2}, \hat{j} - \frac{1}{2})$, or $L(\hat{i} + \frac{1}{2}, \hat{j} + \frac{1}{2}) > L(\hat{i} - \frac{1}{2}, \hat{j} - \frac{1}{2})$. Figure 1 (b) shows an example. All match cells belonging to the same contour form an antichain in a minimal antichain decomposition, and each contour is specified completely by the dominant matches on it[1].

Since parameterized algorithms process the input match-by-match instead of computing the entire prefix-prefix LCS score matrix, it is necessary to pre-process the input strings to obtain lists of match cells. Different approaches exist for this, depending on the assumptions that can be made about the alphabet. Generally, it is necessary to allow less-than/greater-than comparisons in addition to testing for equality (otherwise, $\Omega(mn)$ was shown to be a lower bound [24]). Based on this assumption, we can obtain a set of match lists which give for every character c in x the positions i where $y_i = c$ in $O(n \log n)$ time. These lists usually allow queries for increasing or decreasing values of i and are called occurrence lists or match lists accordingly. The lists are obtained by determining the inverse sorting permutation for y (i.e. a permutation that transforms a sequence which contains all characters from y in sorted order into y). For every character c in x, we can find the head of a list of match positions in time $O(\log n)$ by binary search. For small alphabets, it is possible to pre-process the input in time $O(n \log \sigma)$ to obtain a similar representation (see [15, 25] for discussion). We will denote the result of this preprocessing as follows.

DEFINITION 8. The functions $\mu_i : \mathbb{N} \rightarrow [1 : n] \cup \infty$ for $i \in [1 : m]$ specify the match positions. We have:

- $\mu_i(k) = j, j \neq \infty \Rightarrow x_i = y_j$,

- $\mu_i(k) < \mu_i(k + 1)$ or $\mu_i(k) = \mu_i(k + 1) = \infty$ for all $k \in [1 : n - 1]$.

---

[1]In [21, 22], these contours are called forward contours.

This notation allows storing the match lists using $O(m+n)$ space. We can obtain these functions for arbitrary ordered alphabets in time $O(n \log n)$ by sorting one of the input strings and then using binary search to create the match lists. For small alphabets of size $\sigma < n$, the sorting permutation can be determined in time $O(n \log \sigma)$ by counting character frequencies for all characters contained in y. After this pre-processing step, we can determine $\mu_i(k)$ in $O(1)$ time using $O(m+n)$ storage.

3   The transposition network method

Comparison networks (see e.g. [26]) are a traditional method for studying oblivious algorithms for sorting sequences of numbers. A *comparison network* has n inputs and n outputs, which are connected by an arbitrary number of *comparators*. A comparator has two inputs and two outputs. It compares the input values and returns the larger value at a prescribed output, and the smaller value at the other output. We will draw comparison networks as n wires, where pairs of wires may be connected by comparators that operate on the values passing through the wires. Comparators are usually grouped into a sequence of k *stages*, where each wire is connected to at most one comparator in a single stage. A comparison network is called *transposition network* if all comparators only connect adjacent wires.

Transposition networks allow for another interpretation of the seaweed algorithm. As shown in Figure 4, every mismatch cell behaves like a comparator on the starting points of the seaweeds that enter the cell from the left and the top. The larger value is returned on the right output, and the smaller value is returned on the bottom output. For a match cell, the input values are not compared but just translated top to right and left to bottom. Therefore, we can define a transposition network for every problem instance as follows.

DEFINITION 9. The network $\mathrm{LCSNET}(x, y)$ has $m+n$ diagonal wires. Every mismatch cell $(\hat{\imath}, \hat{\jmath})$ corresponds to a comparator in stage $\hat{\imath} + \hat{\jmath}$ connecting wires $m - \hat{\imath} + \hat{\jmath}$ and $m - \hat{\imath} + \hat{\jmath} + 1$ (see Figure 4). Match cells do not contain comparators.

As comparators in the network correspond to cells in the alignment dag, we choose the convention of drawing the network wires top left to bottom right. Values moving through a cell or comparator can therefore move either down or to the right.

The network $\mathrm{LCSNET}(x, y)$ realizes the seaweed algorithm. The inputs are originally in inversely (in relation to the direction of the comparators) sorted order and trace the seaweed curves on their paths through the transposition network. Note that the direction of the comparators can be determined arbitrarily, as long as it is opposite to the sorting of the input sequence. Another degree of freedom when defining transposition networks lies in the behaviour of comparators for equal inputs. Even though this does not affect the network output, changing the convention of swapping or not swapping equal values can simplify specification of non-oblivious algorithms for computing the output values.

In order to solve the global or semi-local LCS problem for strings x and y using the transposition network method, we have to define appropriate input values for $\mathrm{LCSNET}(x, y)$. In order to obtain the full set of critical points, the inputs are set to the seaweed starting points: input $\hat{j} + m + \frac{1}{2}$ is initialized with $\hat{j}$, $\hat{j} \in \langle -m : n \rangle$. Let the vector O denote the output of the network. If all comparators return the larger input on the bottom output, and the smaller input on the right output, the pairs $(O(\hat{j} + m + \frac{1}{2}), \hat{j} + m)$ with $\hat{j} \in \langle -m : n \rangle$ correspond to the core critical points of the corresponding highest-score matrix. Since there are $O(mn)$ comparators in the transposition network, the resulting algorithm runs in time $O(mn)$.

Using the transposition network method, we can see the connection between semi-local string comparison and existing LCS algorithms is the fact that both approaches compute LCS scores incrementally for prefixes of the input strings: the standard LCS dynamic programming approach computes LCS lengths, and the seaweed algorithm computes implicit highest score matrices for all prefixes of the input strings. When looking at this relationship in more detail, it becomes clear that standard LCS algorithms can be obtained by the transposition network method using input values of only zero or one. A first direct consequence are bit-parallel LCS algorithms [27, 28], which can be obtained by computing the output of the transposition network cell-column by cell-column using bit-vector boolean operations and bit-vector addition. In the remainder of this paper we will show further examples where existing algorithms for comparing two strings globally can be derived from transposition networks, and discuss generalizing them to semi-local string comparison.

4   Sparse semi-local string comparison

We now consider *sparse string comparison*, i.e. string comparison parameterized by the number of matches r in the alignment dag. Hunt and Szymanski [29] proposed an algorithm for sparse string comparison that computes the LCS of two input strings in $O((r + n) \log n)$ time. An extreme case of this is the comparison of permutation strings of length n over the alphabet $\Sigma = [1 : n]$. In this case, only n match cells exist. Tiskin [30] gave an $O(n^{1.5})$ algorithm for semi-local comparison of permutation strings.

Since in sparse string comparison the alignment dag contains few matches,

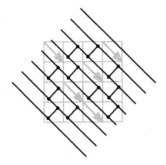

bbbb...b   aaa...aaaa

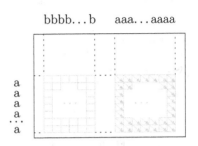

Figure 4. Comparison network of an alignment dag.

Figure 5. Alignment dag for run-length compressed strings.

large rectangular areas of the transposition network have full sets of comparators. These areas will be denoted as follows.

DEFINITION 10. Let network $\mathrm{DIAMOND}(m, n)$ be defined as an LCSNET network which corresponds to a problem instance with no matches. It therefore contains a full set of $m \cdot n$ comparators.

We now give a more general sparse semi-local string comparison algorithm parameterized by the number of matches. We will first show a non-oblivious algorithm to compute the output of DIAMOND networks efficiently, and then propose a technique for evaluating a LCSNET network by partitioning it into smaller DIAMOND networks.

Consider an $m' \times n'$ rectangular area in the alignment dag with only mismatch cells, and the corresponding $\mathrm{DIAMOND}(m', n')$ network. Such an area occurs whenever two substrings over disjoint character sets are compared. The network consists of a full set of $m' \times n'$ comparators and $m' + n'$ wires.[2] If the first m' and the following n' wires are initialized with two pre-sorted sequences of numbers, this network works as a merging network [31]. The problem of merging pre-sorted sequences can be solved non-obliviously in time $O(m' + n')$. However, as the inputs to the DIAMOND network are not necessarily pre-sorted, this is not sufficient.

THEOREM 11. *It is possible to compute the outputs of the DIAMOND network non-obliviously in time $O((m' + n') \log(m' + n'))$ if the inputs are in arbitrary order. Additionally, if the sorting permutation of the inputs is known (but the inputs are still in arbitrary order), the problem can be solved*

---

[2]Note that the opposite case of a rectangular area in the alignment dag which only contains matches is trivially solved in linear time as it corresponds to a transposition network without comparators.

Algorithm 1 Computing the output of DIAMOND(m', n')

---

input : $I[1], \ldots, I[m' + n']$
output : $O[1], \ldots, O[m' + n']$
let $I[L[1]] > I[L[2]] > \ldots > I[L[m' + n']]$ { L is the sorting permutation }
for $(j \in [1 : m' + n'])$ **do** $K[j] \leftarrow$ **false** { K contains the non-free outputs }
$\beta \leftarrow 1$ { β points to the leftmost free output }
for $k = 1, 2, \ldots m' + n'$ { $I[L[k]]$ is the next largest element }
　if $L[k] < \beta + m'$ **then** { Maximum reaches leftmost free output }
　　$O[\beta] \leftarrow I[L[k]]$ { Translate input value to output }
　　$K[\beta] =$ **true** { Mark output as occupied }
　　while $(K[\beta])$ **do** $\beta \leftarrow \beta + 1$
　else { Maximum goes to leftmost output it can reach }
　　$O[L[k] - m'] \leftarrow I[L[k]]$ { Translate input value to output }
　　$K[L[k] - m'] =$ **true** { Mark output as occupied }
　end if
end while

---

in $O(m' + n')$ time, as the factor of $\log(m' + n')$ only comes from the initial sorting step.

Proof. To non-obliviously compute the output of DIAMOND(m', n'), consider the path that the largest input takes through the network. If the largest input enters the network on wire j, all comparators it passes will return it as the larger element, which means that it will reach the leftmost output possible. We then proceed through the remaining inputs in descending order, determining for every input the leftmost output it can reach, considering that some outputs have already been occupied by larger values. Any current value that enters the comparison network on a wire j that is less than m' wires ahead of the first free output will be translated to the first (leftmost) available output. If the current value enters the network more than m' wires to the right of the first available output, it can only pass through m' comparators and will therefore reach output $j - m'$. The free outputs are indicated by a Boolean array K, where occupied outputs are marked with a value of true. Since we proceed through the input values in descending order, this yields the same output as direct evaluation of the transposition network. The entire algorithm is shown in Algorithm 1. ∎

Using Algorithm 1, we obtain an improved algorithm for sparse semi-local comparison. For simplicity assume that both strings are of length n and (w.l.o.g.) that n is a power of 2.

THEOREM 12. *After pre-processing the input strings for obtaining match* *lists, the problem of semi-local string comparison can be solved in $O(n\sqrt{r})$ time.*

Proof. We first find the sorting permutations of the input strings. This is possible in time $O((m+n)\log\min(\sigma, \max(m,n)))$, similar to obtaining μ_i in Section 2.

After this pre-processing, we partition the alignment dag into blocks using a recursive quadtree scheme. Consider processing such a block of size $w \times w$. Let this block correspond to comparing substrings $x_k \ldots x_{k+w-1}$ and $y_l \ldots y_{l+w-1}$. As an input for each such block, we have the sorting permutations of the two corresponding substrings, the input values for the transposition network corresponding to the block, and also the sorting permutation for these input values. For each block, we obtain the output values of its transposition network and their sorting permutation as follows.

For a $w \times w$ block, we can count the number of matches in it in time $O(w)$ by linear search in the sorting permutations of the corresponding substrings. Whenever we find a block that does not contain any matches, we stop partitioning and use Algorithm 1 to compute the outputs of the corresponding comparison network. Otherwise, we continue to partition until we obtain a 1×1 block that only consists of a single match.

A 1×1 leaf block consisting of a single match can be processed trivially in constant time. Due to Theorem 11, we can compute the outputs for a $w \times w$ mismatch block in $O(w)$ time when the sorting permutation is known for the inputs. The sorting permutation for the root block of the quadtree is known, since the root of the quadtree corresponds to the full alignment dag, and the inputs to its transposition network form a sequence sorted in reverse. For all other blocks, we keep track of the sorting permutation of both its input and output elements. For every output we can trace the input it came from before executing Algorithm 1 and therefore know the permutation that was performed by the transposition network within the block. Knowing this permutation and the sorting permutation of the inputs allows to establish the sorting permutation of the outputs in time $O(w)$.

To summarize, given the input values and their sorting permutation for every leaf block of the quadtree recursion, we can compute the output values and their sorting permutation in time $O(w)$. All non-leaf blocks are partitioned into four sub-blocks of size $w/2 \times w/2$. The inputs and their sorting permutation are split and used to recursively process the sub-blocks. We can then establish the sorting permutation of the outputs for the entire block in linear time by merging. To compute the outputs of any intermediate block we therefore need time $O(w)$ in addition to the time necessary for recursively processing the sub-blocks.

Consider the top $\log_4 r$ levels of the quadtree. In each subsequent level, the number of blocks increases by at most a factor of four, and the block size decreases by a factor of two. Therefore, this part of the quadtree is dominated by level $\log_4 r$ which contains at most r blocks, each of size $n/\sqrt{r}$. The total work required on this part of the tree is therefore $O(r \cdot n/\sqrt{r}) = O(n\sqrt{r})$.

The remaining levels of the quadtree can each have at most r blocks that still contain matches. The block size in each level still decreases by a factor of two. Therefore, this part of the quadtree is also dominated by level $\log_4 r$ and requires the same asymptotic amount of work. The overall time for the algorithm is therefore bounded by $\sum_{j=0}^{\log_4 r} O(n/2^j \cdot 4^j) + \sum_{j=\log_4 r+1}^{\log_4 n} O(n/2^j \cdot r) = O(n\sqrt{r})$. The resulting algorithm has running time $O(n\sqrt{r})$, and thus provides a smooth transition between the dense case ($r = n^2$, running time $O(n^2)$) and the permutation case ($r = n$, running time $O(n^{1.5})$). ∎

5   Semi-local comparison of run-length compressed strings

Another straightforward application of Algorithm 1 is comparing run-length compressed strings [32]. In this compression method, a run of repeating characters is encoded by a single character together with the number of repetitions. A run-length encoded string $X = X_1 X_2 X_3 \ldots X_{\overline{m}}$ consists of $\overline{m}$ character runs X_j of lengths $|X_j|$. The length of the full string is therefore $m = \sum_{j=1\ldots\overline{m}} |X_j|$. When constructing the alignment dag for comparing two run-length compressed strings $X = X_1 X_2 X_3 \ldots X_{\overline{m}}$ and $Y = Y_1 Y_2 Y_3 \ldots Y_{\overline{n}}$, rectangular areas without matches occur when character runs in X and Y mismatch. Analogously, large rectangular areas with containing only match cells occur if the characters do match (see Figure 5). Using the comparison network method and Algorithm 1, these rectangular areas can be processed in cost proportional to their perimeter. Given two input strings with uncompressed lengths m and n, and compressed lengths $\overline{m}$ and $\overline{n}$, this method results in an algorithm for semi-local comparison which has cost $\sum_{i\in[1:\overline{m}], j\in[1:\overline{n}]} O(|X_i| + |Y_j|) = O(\overline{m}n + m\overline{n})$. This is as good as the result from [33], additionally solving the more general problem of semi-local string comparison of run-length compressed strings.

6   High similarity and dissimilarity string comparison

In Section 4 we described an efficient algorithm for semi-local string comparison, parameterized by the overall number of matches. We now describe an application of the transposition network method to designing algorithms that are parameterized by the LCS length p of the input strings or their LCS distance $k = n - p$. Such parametrization provides efficient algorithms when the corresponding parameter is low, i.e. when the strings are highly dissimilar or highly similar.

In [29], matches are processed row by row to establish which antichain they belong to. Apostolico and Guerra improved this algorithm by avoiding the need to consider non-dominant matches [34] (see Section 2), and changing the order in which the match cells are processed. This allows to obtain an algorithm that is parameterized by the length of the LCS. Further, there have been various extensions to this approach, which improve the running time by either using different data structures [35] or narrowing the area in which to search for dominant matches hence giving algorithms which are efficient both when the LCS of the two strings is long or short [21]. In this paper, we will show how the transposition network method can be used to match these algorithms for global LCS computation. For semi-local alignment, we achieve a running time of $O(np)$, which is efficient for dissimilar strings.

We will now show the connection between the antichain decomposition of the set of match cells and the transposition network method. Consider an LCSNET network with the following input values: The first m wires (i.e. the inputs on left hand side of the alignment dag) are initialized with ones, and the following n wires (i.e. the inputs at the top of the alignment dag) are filled with zeros. On all comparators, smaller values are returned at the bottom output. We will refer to this specific transposition network setup as $LCSNET(x,y)$ with 0/1 inputs. Using only zeros and ones as inputs to LCSNET(x,y) corresponds to tracing seaweeds anonymously, only distinguishing between those seaweeds that start at the top and those seaweeds that start at the left. The 0-1 transposition network approach allows to understand previous results for parameterized LCS computation in terms of transposition networks, and helps to extend some of these to semi-local string comparison.

COROLLARY 13. *In* LCSNET(x,y) *with 0/1 inputs as described above, let p be the number of ones reaching output wires below $m+1$ (i.e. the bottom of the alignment dag). This number is equal to the number of zeros reaching an output wire above m (i.e. the right side of the alignment dag), and $LLCS(x,y) = p$.*

Proof. From Theorem 5, we know that $LLCS(x,y) = n - d$, where d is the number of seaweeds that start at the top and end at the bottom of the alignment dag. The number of zeros ending up at the bottom is therefore equal to d, and the number of ones ending up at the bottom is equal to $n - d = LLCS(x,y)$. Since the transposition network outputs a permutation of the input, and since we have n input zeros, $n - d$ zeros must end up at the right.   ∎

We will now look at the behaviour of LCSNET(x, y) with $0/1$ inputs in
more detail. In order to be able to trace paths of individual values, we must
specify the behaviour of the comparators for equal input values (note that
changing this specification does not change the output of LCSNET(x, y)).
Assume that comparators in LCSNET(x, y) swap their input values if these
are equal. If the alignment dag contains only mismatch cells and therefore
a full set of comparators, all ones move from the left to the right, and all
zeros move from the top to the bottom. When introducing a match cell and
hence removing a comparator, the zero that enters the match cell at the
top is translated to the right, and the value of one entering the match cell
at the left is translated to the bottom. We trace these two values further:
as identical values are swapped by convention, both the one (and equally
the zero) will not change direction of movement and be passed on vertically
(horizontally in case of the zero) through all comparators. We will refer
to ones which move downwards and to zeros which move to the right as
stray. Stray values only change direction again when they either encounter
a match cell or another stray value. If two stray values enter the same cell,
they leave this cell in the original directions, the one moving rightwards,
and the zero moving downwards. This happens independently of whether
this cell contains a match: in a match cell, no comparison is performed,
the stray zero is returned at the bottom and the stray one is returned at
the right. In a mismatch cell, the zero is also returned at the bottom since
it is the smaller value. Therefore, two stray values always return to their
original direction of movement when meeting in the same cell. Another
observation is that any cell which has exactly one stray input value must
have equal inputs. If such a cell is a match cell, the stray input value returns
to its original direction of movement, and the other input becomes stray. If
the cell does not contain a match, the inputs are exchanged by convention,
and the stray value remains stray. To summarize, stray values caused by
a match cell will start a row (stray zeros) or column (stray ones) of cells
which output stray values. This row or column only ends when meeting
another column or row of cells which output stray values.

Figure 6 shows an example of the LCSNET(x, y) with $0/1$ inputs for the
problem instance shown in Figure 1 on page 187. It seems intuitive from
this figure that the stray zeros and ones trace contours in L.

THEOREM 14. *A cell belongs to a contour in the matrix of prefix-prefix
LCS lengths L iff it has at least one stray value as an input or output.*

Proof. This follows from Corollary 13 by induction on the number of
contours. If L has no contours, no match cells can exist. If there is exactly

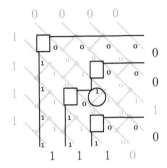

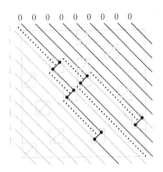

Figure 6. LCSNET(x, y) with 0/1 inputs.

Figure 7. Comparing highly similar strings.

one contour in L, all match cells must belong to this contour, and the contour splits the set of cells into two parts of mismatch cells. Consider the set of mismatch cells to the top/left of the contour. All cells in this set have zeros as their top input and ones as their left input since these are either the input values to the transposition network, or have been translated through the previous mismatch cells as shown in case (e) of Figure 8. All dominant matches on the contour must have a zero as their top input and a one as their left input as well, since they must be at the right and below a case (e) mismatch cell, or equivalently at the top or left of the alignment dag. Dominant match cells output a stray zero on the right and a stray one on the bottom (see case (d) in Figure 8). Any cell that has a stray zero as its left input and a zero as its top input must be to the right of a match. As there is only one contour the cell cannot be below another match and therefore L will increase vertically in this cell since the prefix-prefix LCS can be extended by the first match to the left. Symmetrically, this is true for any cell with a stray and a none-stray one as its inputs (see cases (a) and (b) in Figure 8). In the only remaining case, two stray values meet in the same cell $(\hat{i}, \hat{j})$ (case (c) in Figure 8). In this case, the prefix-prefix LCS could either be extended by using the matches above $(\hat{i}, \hat{j})$ or by using the matches to the left of $(\hat{i}, \hat{j})$, but not by using both since they are incomparable under the $\prec$ ordering (and no path containing one of each of those matches exists in the alignment dag). Now consider the cells immediately to the right or below the contour. These cells cannot be to the right or below dominant matches (otherwise they would belong to the contour). Therefore, these cells must all have non-stray inputs (i.e. a zero at the top input and a one at the left input), since cells on the horizontal contour output zeros on the

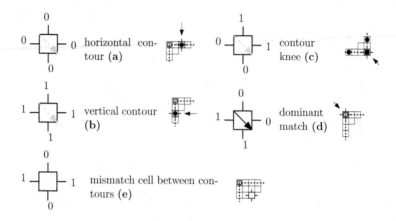

Figure 8. Interpreting 0-1 transposition network cells and their inputs as
contours.

bottom, cells on the vertical contour output ones at the right, and contour
knees output a zero on the bottom and a one at the right output. As all the
cells immediately neighbouring the contour to the right or below must be
mismatch cells (only one contour exists ⇒ all match cells are on it), they
all belong to case (e) in Figure 8 and in consequence all cells below or to
the right of them as well. Therefore, Theorem 14 is true in the case where
only one contour exists. Furthermore, all additional contours must either
have case (e) cells on top and to their left, or border directly on another
contour. Cell contours output non-stray values on the right/the bottom if
they have non-stray inputs. Therefore, Theorem 14 is also true for more
than one contour.   ∎

The resulting algorithm is for computing output of $LCSNET(x, y)$ with $0/1$
inputs is equivalent to [34], giving a running time of $O(mn \log p)$ which
can be improved to $O(m \log n + d \log(mn/d))$ using the finger searching
technique [36].

   Consider the problem of comparing two strings that are highly similar.
Myers [18] proposed an algorithm to compare strings in time $O(ne)$, where
e is the edit distance between the strings. The idea behind this algorithm
is to incrementally extend only the longest paths in the alignment dag until the LCS is found. A similar algorithm can be obtained by using 0-1
transposition networks as follows.

   If the two input strings are identical, no comparators exist on the main
diagonal of alignment dag cells, i.e. between transposition network wires m

and $m + 1$. This means that no ones can get to the right hand side, and no zeros can get to the bottom of the alignment dag. We can look at this as two streams of zeros and ones, and do not need to evaluate comparisons within a single stream of zeros or ones. The only comparators which can possibly swap inputs are the ones between streams. If a comparator occurs between two streams, the inputs will only be swapped if the zero is input from the top, i.e. we can restrict our attention to the upper boundaries of streams of ones. Figure 7 shows an example. The comparators drawn in black are those between streams of zeros and ones which must swap their inputs.

DEFINITION 15. Let a *1-0 boundary* in stage s of LCSNET(x, y) with $0/1$ inputs be defined as any location in this stage where two adjacent wires l and $l + 1$ carry values one and zero respectively.

COROLLARY 16. *The number of 1-0 boundaries in any stage of the transposition network is dominated by* $k + 1 = n - p + 1$.

Proof. By induction: Assume $m = n = 1$. The transposition network has two wires which are initialized with a zero and a one. Therefore, the number of 1-0 boundaries must be less or equal than 1. The LCS distance k can be 0 or 1. Increasing m or n by one adds another row or column of comparators to the transposition network. Consider the case of adding a column of comparators (i.e. increasing n by one). Each 1 which is output at the right hand side can only cause one 1-0 boundary. Furthermore, ones do not move downwards in comparisons. Therefore, a new 1-0 boundary can only be created if a value of 1 from the left hand side reaches the right hand side, which means that the number of 1-0 boundaries cannot increase by one in this case without also increasing k by one. However, k cannot increase by more than one, since maximally a single value of 1 reaches the right hand side. Symmetrically, when increasing m by one, we add a row of comparators at the bottom. If we have k zeros at the bottom, each of these zeros can only be part of a single 1-0 boundary. We can only gain a single 0 on the bottom by increasing m by one, in which case also k increases. Therefore $k + 1$ always dominates the number of 1-0 boundaries.   ∎

Using this insight, the LLCS of two strings x and y with $|x| = |y| = n$ can be computed in time $O(nk)$. This is done by tracing the intersections of the 1-0 boundaries with the $2n - 1$ antidiagonals of the alignment dag, as this is the only place where change can occur. By Corollary 16, we know a bound for the number of 1-0 boundaries. At each intersection of a 1-0 boundary with an antidiagonal, the corresponding characters in x and y must be compared to check whether a comparator exists. This can be done in constant time, and since there are $2n - 1$ antidiagonals we get the claimed running time. Note that this algorithm does not require any pre-processing to obtain match lists.

COROLLARY 17. *All dominant matches must be on a 1-0 boundary in the transposition network.*

Proof. This follows immediately from Theorem 14.   ∎

Corollary 17 allows to narrow down the area in which to search for dominant matches, and can be used to extend Algorithm [34] to achieve running time $O(kp)$, similarly to [21].

THEOREM 18. *The implicit highest-score matrix for comparing two strings of length n can be computed in time $O(np)$.*

Proof. Using the 0-1 transposition network, we are able to determine for every match cell whether it is dominant or non-dominant, as well as for every mismatch cell whether it is part of a contour. Looking at this in the more general setting of semi-local string comparison where we need to trace all seaweeds individually, we can still see that non-trivial comparisons between seaweeds can only occur when the cell is actually part of a contour. Cells outside the contours are always mismatch cells which compare an input originating at the left hand side of the alignment dag to an input originating at the top of the alignment dag. Therefore all the comparators in these cells can be replaced by swap operations (i.e. they contain seaweed crossings).

Given all dominant matches on a contour and the values on all transposition network wires before they intersect the contour, we can compute the values on all wires of transposition network after the intersection in time which is linear in the length of the contour. As all comparators between contours perform swap operations, we can also compute the permutation of values performed between two contours in time linear in the length of the longer contour.

It is possible to compute the set of all k-dominant matches with $k \in [1 : p]$ in $O(np)$ time. We can use Algorithm [34][3] for this. Knowing the dominant matches in every antichain, we can trace its complete contour in time linear in its length. No contour can have length l longer than $2n$, and there are exactly $p = \mathrm{LLCS}(x, y)$ contours. Further, we can obtain the inputs and outputs of all cells in a contour of length l in time $O(l)$ with $l \leq 2n$. Therefore, the worst case running time of our algorithm for semi-local string comparison is bounded by $O(np)$.   ∎

---

[3] A practical algorithm for computing a list of dominant matches is described in [28]

7 Conclusions

In this paper, we have presented a new method of solving the semi-local string comparison problem using transposition networks. This method provides a unified view of different string comparison algorithms, and allows to obtain efficient algorithms for global string comparison which have the same complexity as the best known algorithms. Furthermore, we have obtained new algorithms for sparse semi-local string comparison, high similarity and dissimilarity string comparison, as well as semi-local comparison of run-length compressed strings. In a separate paper, we will show that it is possible to implement the algorithms for semi-local string comparison efficiently for an application to LCS-filtered dot-plots [37]. We conclude that the transposition network method is a very general and flexible way of understanding and improving different string comparison algorithms.

BIBLIOGRAPHY

[1] Levenshtein, V.I.: Binary codes capable of correcting deletions, insertions and reversals,. Sov. Phys. Dokl. **6** (1966) 707–710

[2] Hirschberg, D.S.: Serial computation of Levenshtein distances. In Apostolico, A., Galil, Z., eds.: Pattern Matching Algorithms. Oxford University Press (1997) 123–141

[3] Navarro, G.: A guided tour to approximate string matching. ACM Computing Surveys **33**(1) (2001) 31–88

[4] Schmidt, J.P.: All highest scoring paths in weighted grid graphs and their application to finding all approximate repeats in strings. SIAM Journal on Computing **27**(4) (1998) 972–992

[5] Alves, C., Caceres, E., Song, S.: An all-substrings common subsequence algorithm. Discrete Applied Mathematics **156**(7) (April 2008) 1025–1035

[6] Tiskin, A.: Semi-local longest common subsequences in subquadratic time. Journal of Discrete Algorithms **6**(4) (2008) 570–581

[7] Boasson, L., Cégielski, P., Guessarian, I., Matiyasevich, Y.: Window-accumulated subsequence matching problem is linear. Ann. Pure Appl. Logic **113**(1-3) (2001) 59–80

[8] Apostolico, A., Atallah, M.J., Larmore, L.L., McFaddin, S.: Efficient parallel algorithms for string editing and related problems. SIAM J. Comput. **19**(5) (1990) 968–988

[9] Alves, C.E.R., Cáceres, E.N., Song, S.W.: A coarse-grained parallel algorithm for the all-substrings longest common subsequence problem. Algorithmica **45**(3) (2006) 301–335

[10] Krusche, P., Tiskin., A.: Efficient parallel string comparison. In: ParCo. Volume 38 of NIC Series., John von Neumann Institute for Computing (2007) 193–200

[11] Tiskin, A.: Semi-local string comparison: Algorithmic techniques and applications. Mathematics in Computer Science **1**(4) (2008) 571–603 See also arXiv: 0707.3619.

[12] Wagner, R.A., Fischer, M.J.: The string-to-string correction problem. Journal of the ACM **21**(1) (1974) 168–173

[13] Burkard, R.E., Klinz, B., Rudolf, R.: Perspectives of Monge properties in optimization. Discrete Applied Mathematics **70**(2) (1996) 95–161

[14] Tiskin, A.: Efficient representation and parallel computation of string-substring longest common subsequences. In: Proceedings of ParCo. Volume 33 of NIC Series., John von Neumann Institute for Computing (2005) 827–834

[15] Hirschberg, D.S.: Algorithms for the longest common subsequence problem. Journal of the ACM **24**(4) (1977) 664–675
[16] Nakatsu, N., Kambayashi, Y., Yajima, S.: A longest common subsequence algorithm suitable for similar text strings. Acta Informatica **18**(2) (1982) 171–179
[17] Ukkonen, E.: Algorithms for approximate string matching. Information and Control **64**(1–3) (1985) 100–118
[18] Myers, E.W.: An $O(ND)$ difference algorithm and its variations. Algorithmica **1**(1) (1986) 251–266
[19] Kumar, S.K., Rangan, C.P.: A linear space algorithm for the LCS problem. Acta Informatica **24**(3) (1987) 353–362
[20] Wu, S., Manber, U., Myers, G.: An $O(NP)$ sequence comparison algorithm. Information Processing Letters **35**(6) (1990) 317–323
[21] Rick, C.: A new flexible algorithm for the longest common subsequence problem. Nordic Journal of Computing **2**(4) (1995) 444–461
[22] Rick, C.: Simple and fast linear space computation of longest common subsequences. Information Processing Letters **75**(6) (2000) 275–281
[23] Dilworth, R.P.: A decomposition theorem for partially ordered sets. Ann. Math. **51** (1950) 161–166
[24] Aho, A.V., Hirschberg, D.S., Ullman, J.D.: Bounds on the complexity of the longest common subsequence problem. Journal of the ACM **23** (1976) 1–12
[25] Apostolico, A.: String editing and longest common subsequences. In: Handbook of Formal Languages. Volume 2. Springer-Verlag (1997) 361–398
[26] Cormen, T.H., Leiserson, C.E., Rivest, R.L., Stein, C.: Introduction to Algorithms, Second Edition. The MIT Press and McGraw-Hill Book Company (2001)
[27] Crochemore, M., Iliopoulos, C.S., Pinzon, Y.J., Reid, J.F.: A fast and practical bit-vector algorithm for the longest common subsequence problem. Information Processing Letters **80**(6) (December 2001) 279–285
[28] Crochemore, M., Iliopoulos, C.S., Pinzon, Y.J.: Speeding-up Hirschberg and Hunt–Szymanski algorithms for the LCS problem. Fundamenta Informaticae **56**(1–2) (2003) 89–103
[29] Hunt, J.W., Szymanski, T.G.: A fast algorithm for computing longest common subsequences. Communications of the ACM **20**(5) (May 1977) 350–353
[30] Tiskin, A.: Longest common subsequences in permutations and maximum cliques in circle graphs. In: Proceedings of CPM, vol. 4009 of Lecture Notes in Computer Science. (2006) 271–282
[31] Munter, E.A.: U.S. Patent 5,216,420 (June 1993)
[32] Apostolico, A., Landau, G.M., Skiena, S.: Matching for run-length encoded strings. J. Complexity **15**(1) (1999) 4–16
[33] Bunke, H., Csirik, J.: An algorithm for matching run-length coded strings. Computing **50**(4) (1993) 297–314
[34] Apostolico, A., Guerra, C.: The longest common subsequence problem revisited. Algorithmica **2**(1) (March 1987) 315–336
[35] Eppstein, D., Galil, Z., Giancarlo, R., Italiano, G.F.: Sparse dynamic programming i: linear cost functions. Journal of the ACM **39**(3) (1992) 519–545
[36] Brodal: Finger search trees. In Mehta, Sahni, eds.: Handbook of Data Structures and Applications. Chapman& Hall/CRC (2005)
[37] Maizel, J.V., Lenk, R.P.: Enhanced graphic matrix analysis of nucleic acid and protein sequences. Proceedings of the National Academy of Sciences of the USA **78**(12) (1981) 7665–7669

Peter Krusche
Department of Computer Science
University of Warwick

Coventry CV4 7AL, UK
Email: peter@dcs.warwick.ac.uk

Alexander Tiskin
Department of Computer Science
University of Warwick
Coventry CV4 7AL, UK
Email: tiskin@dcs.warwick.ac.uk

www.ingramcontent.com/pod-product-compliance
Lightning Source LLC
LaVergne TN
LVHW012329060326
832902LV00011B/1796